Advancing Research on Understanding Environmental Effects of UV Filters from Sunscreens

Anne Frances Johnson, *Rapporteur*

Ocean Studies Board

Board on Environmental Studies and Toxicology

Board on Health Sciences Policy

Division on Earth and Life Studies

Health and Medicine Division

Proceedings of a Workshop

THE NATIONAL ACADEMIES PRESS 500 Fifth Street, NW Washington, DC 20001

This study was supported by a contract between the National Academy of Sciences and the U.S. Environmental Protection Agency. Any opinions, findings, conclusions, or recommendations expressed in this publication do not necessarily reflect the views of any organization or agency that provided support for the project.

International Standard Book Number-13: 978-0-309-69585-5
International Standard Book Number-10: 0-309-69585-6
Digital Object Identifier: https://doi.org/10.17226/26775

This publication is available from the National Academies Press, 500 Fifth Street, NW, Keck 360, Washington, DC 20001; (800) 624-6242 or (202) 334-3313; http://www.nap.edu.

Copyright 2023 by the National Academy of Sciences. All rights reserved.

Printed in the United States of America

Suggested citation: National Academies of Sciences, Engineering, and Medicine. 2023. *Advancing Research on Understanding Environmental Effects of UV Filters from Sunscreens: Proceedings of a Workshop*. Washington, DC: The National Academies Press. https://doi.org/10.17226/26775.

The **National Academy of Sciences** was established in 1863 by an Act of Congress, signed by President Lincoln, as a private, nongovernmental institution to advise the nation on issues related to science and technology. Members are elected by their peers for outstanding contributions to research. Dr. Marcia McNutt is president.

The **National Academy of Engineering** was established in 1964 under the charter of the National Academy of Sciences to bring the practices of engineering to advising the nation. Members are elected by their peers for extraordinary contributions to engineering. Dr. John L. Anderson is president.

The **National Academy of Medicine** (formerly the Institute of Medicine) was established in 1970 under the charter of the National Academy of Sciences to advise the nation on medical and health issues. Members are elected by their peers for distinguished contributions to medicine and health. Dr. Victor J. Dzau is president.

The three Academies work together as the **National Academies of Sciences, Engineering, and Medicine** to provide independent, objective analysis and advice to the nation and conduct other activities to solve complex problems and inform public policy decisions. The National Academies also encourage education and research, recognize outstanding contributions to knowledge, and increase public understanding in matters of science, engineering, and medicine.

Learn more about the National Academies of Sciences, Engineering, and Medicine at **www.nationalacademies.org**.

Consensus Study Reports published by the National Academies of Sciences, Engineering, and Medicine document the evidence-based consensus on the study's statement of task by an authoring committee of experts. Reports typically include findings, conclusions, and recommendations based on information gathered by the committee and the committee's deliberations. Each report has been subjected to a rigorous and independent peer-review process and it represents the position of the National Academies on the statement of task.

Proceedings published by the National Academies of Sciences, Engineering, and Medicine chronicle the presentations and discussions at a workshop, symposium, or other event convened by the National Academies. The statements and opinions contained in proceedings are those of the participants and are not endorsed by other participants, the planning committee, or the National Academies.

Rapid Expert Consultations published by the National Academies of Sciences, Engineering, and Medicine are authored by subject-matter experts on narrowly focused topics that can be supported by a body of evidence. The discussions contained in rapid expert consultations are considered those of the authors and do not contain policy recommendations. Rapid expert consultations are reviewed by the institution before release.

For information about other products and activities of the National Academies, please visit www.nationalacademies.org/about/whatwedo.

PLANNING COMMITTEE FOR WORKSHOP TO ADVANCE RESEARCH ON UNDERSTANDING ENVIRONMENTAL EFFECTS OF UV FILTERS FROM SUNSCREENS

CHARLES MENZIE (*Chair*), Exponent, Inc
SCOTT BELANGER, Procter and Gamble (retired)
CHRISTOPHER P. HIGGINS, Colorado School of Mines
REBECCA D. KLAPER, University of Wisconsin, Milwaukee
CARYS MITCHELMORE, University of Maryland
ROBERT RICHMOND, University of Hawaii, Manoa
EMMA J. ROSI, Cary Institute
PAUL K. WESTERHOFF, Arizona State University
CHERYL WOODLEY, NOAA

National Academies of Sciences, Engineering, and Medicine Staff

SUSAN ROBERTS, Director
EMILY TWIGG, Senior Program Officer (until May 2023)
LEIGHANN MARTIN, Research Associate
ERIK YANISKO, Program Assistant

BOARD ON ENVIRONMENTAL STUDIES AND TOXICOLOGY

FRANK W. DAVIS (*Chair*), University of California, Santa Barbara
DANA BOYD BARR, Emory University
ANN M. BARTUSKA, U.S. Department of Agriculture (retired)
GERMAINE M. BUCK LOUIS, George Mason University
FRANCESCA DOMINICI, Harvard University
R. JEFFREY LEWIS, ExxonMobil Biomedical Sciences, Inc.
MARIE LYNN MIRANDA, University of Notre Dame
REZA J. RASOULPOUR, Corteva Agriscience
JOSHUA TEWKSBURY, Smithsonian Tropical Research Institute
SACOBY M. WILSON, University of Maryland, College Park
TRACEY JEAN WOODRUFF, University of California, San Francisco

National Academies of Sciences, Engineering, and Medicine Staff

CLIFFORD S. DUKE, Director
RAYMOND WASSEL, Scholar
KATHRYN GUYTON, Senior Program Officer
NATALIE ARMSTRONG, Associate Program Officer
ANTHONY DEPINTO, Associate Program Officer
LAURA LLANOS, Finance Business Partner
LESLIE BEAUCHAMP, Senior Program Assistant
THOMASINA LYLES, Senior Program Assistant
KATHERINE KANE, Program Assistant

BOARD ON HEALTH SCIENCES POLICY

SHARON TERRY (*Chair*), Genetic Alliance
DAVID BLAZES, Bill and Melinda Gates Foundation
ARAVINDA CHAKRAVARTI, New York University
AMANDER CLARK, University of California, Los Angeles
STEVE K. GALSON, Amgen Incorporated (retired)
M. EHSAN HOQUE, University of Rochester
FRANCES E. JENSEN, University of Pennsylvania
FRANK R. LIN, Johns Hopkins Medical Institutions
SUZET. M MCKINNEY, Sterling Bay
DIETRAM A. SCHEUFELE, University of Wisconsin, Madison
MATTHEW K. WYNIA, University of Colorado
PATRICIA ZETTLER, The Ohio State University

National Academies of Sciences, Engineering, and Medicine Staff

CLARE STROUD, Senior Board Director
SARAH BEACHY, Senior Program Officer
KATHERINE BOWMAN, Senior Program Officer
LISA BROWN, Senior Program Officer
AUTUMN DOWNEY, Senior Program Officer
REBECCA ENGLISH, Senior Program Officer
SHEENA POSEY NORRIS, Senior Program Officer
CAROLYN SHORE, Senior Program Officer
RAYANE SILVA-CURRAN, Senior Program Officer
SCOTT WOLLEK, Senior Program Officer
ANDREW MARCH, Program Officer
CHANEL MATNEY, Program Officer
SHALINI SINGARAVELU, Program Officer
TEQUAM WORKU, Program Officer
OLIVIA YOST, Program Officer
KATHRYN ASALONE, Associate Program Officer
KELSEY BABIK, Associate Program Officer
EVA CHILDERS, Associate Program Officer
MEREDITH HACKMAN, Associate Program Officer
MATTHEW MASIELLO, Associate Program Officer
AARON RESNICK, Associate Program Officer
SAMANTHA SCHUMM, Associate Program Officer
MAYA THIRKILL, Associate Program Officer
MICHAEL BERRIOS, Research Associate
MARGARET MCCARTHY, Research Associate
EMILY MCDOWELL, Research Associate
NOAH ONTJES, Research Associate
LYDIA TEFERRA, Research Associate
ASHLEY BOLOGNA, Senior Program Assistant

VICTORIA CHEGE, Senior Program Assistant
APARNA CHERAN, Senior Program Assistant
MELVIN JOPPY, Senior Program Assistant
EDEN NELEMAN, Senior Program Assistant
GINA STROHBACH, Program Coordinator

OCEAN STUDIES BOARD

CLAUDIA BENITEZ-NELSON (*Chair*), University of South Carolina, Columbia
MARK R. ABBOTT, Woods Hole Oceanographic Institution
ROSANNA 'ANOLANI ALEGADO, University of Hawai'i, Manoa
CAROL ARNOSTI, University of North Carolina at Chapel Hill
AMY BOWER, Woods Hole Oceanographic Institution
LISA M. CAMPBELL, Duke University
THOMAS S. CHANCE, ASV Global, LLC (retired)
DANIEL COSTA, University of California, Santa Cruz
JOHN R. DELANEY, University of Washington (retired)
TIMOTHY GALLAUDET, Ocean STL Consulting, LLC
SCOTT GLENN, Rutgers University
MARCIA ISAKSON, The University of Texas at Austin
LEKELIA JENKINS, Arizona State University, Tempe
NANCY KNOWLTON (NAS), Smithsonian Institution (retired)
ANTHONY MACDONALD, Monmouth University
GALEN MCKINLEY, Columbia University
THOMAS J. MILLER, University of Maryland, Solomons
S. BRADLEY MORAN, University of Alaska Fairbanks
RUTH PERRY, Shell Exploration & Production Company
DEAN ROEMMICH, Scripps Institute of Oceanography (retired)
JAMES SANCHIRICO, University of California, Davis
MARK J. SPALDING, The Ocean Foundation
PAUL WILLIAMS, Squamish Indian Tribe

National Academies of Sciences, Engineering, and Medicine Staff

SUSAN ROBERTS, Director
STACEE KARRAS, Senior Program Officer
KELLY OSKVIG, Senior Program Officer
EMILY TWIGG, Senior Program Officer (until May 2023)
CAROLINE BELL, Associate Program Officer
THANH NGUYEN, Financial Business Partner
LEIGHANN MARTIN, Research Associate
ERIK YANISKO, Program Assistant
SAFAH WYNE, Program Assistant

Reviewers

These Proceedings of a Workshop to Advance Research on Understanding Environmental Effects of UV Filters in Sunscreens was reviewed in draft form by individuals chosen for their diverse perspectives and technical expertise. The purpose of this independent review is to provide candid and critical comments that will assist the National Academies of Sciences, Engineering, and Medicine in making each published proceedings as sound as possible and to ensure that it meets the institutional standards for quality, objectivity, evidence, and responsiveness to the charge. The review comments and draft manuscript remain confidential to protect the integrity of the process.

We thank the following individuals for their review of these proceedings:

SAMANTHA KORETSKY, National Academies of Sciences, Engineering, and Medicine
ROBERT RICHMOND, University of Hawai'i at Mānoa

Although the reviewers listed above provided many constructive comments and suggestions, they were not asked to endorse the content of the proceedings nor did they see the final draft before its release. The review of these proceedings was overseen by **JUDITH E. MCDOWELL**, Woods Hole Oceanographic Institute. She was responsible for making certain that an independent examination of these proceedings was carried out in accordance with standards of the National Academies and that all review comments were carefully considered. Responsibility for the final content rests entirely with the rapporteur(s) and the National Academies.

Contents

INTRODUCTION AND CONTEXT 1
    Workshop Context, 1
    Management Context, 3

UV FILTER CHEMISTRY FOR ACCURATE DOSE–RESPONSE RELATIONSHIPS ... 5
    The Environmental Fate of UV Filters, 5
    Analytical Challenges in Quantifying Organic UV Filters, 6
    Panel Discussion, 8
    Breakout Discussions, 11

STANDARDIZING APPROACHES FOR TOXICITY TESTING 14
    EPA Perspective: The Importance of Standardized Toxicological Methods for
      Aquatic Organisms, 14
    Lightning Talks: Methods for Coral Ecotoxicology, 15
    Panel Discussion, 20
    Breakout Discussions, 24

CLOSING REMARKS 27

APPENDIXES

A Statement of Task 29

B Workshop Agenda 31

C Biosketches for Workshop Planning Committee Members 35

Introduction and Context

Sunscreens and a variety of other products contain chemical ultraviolet (UV) filters that absorb or block the sun's radiation and thereby help mitigate harms to human skin from the sun. These UV filters and the chemicals with which they are mixed can enter bodies of water directly, such as when people swim after applying sunscreen, or indirectly, such as through wastewater effluent carrying compounds that are rinsed off when bathing or showering. The 17 UV filters currently in use in the United States exhibit a wide range of properties in terms of biodegradability and toxicity, raising concerns about the potential for harmful impacts on the environment and in particular, on aquatic and marine organisms.

The 2022 National Academies of Sciences, Engineering, and Medicine (NASEM) report *Review of Fate, Exposure, and Effects of Sunscreens in Aquatic Environments and Implications for Sunscreen Usage and Human Health*[1] (referred to as "2022 report" throughout these proceedings) called on the U.S. Environmental Protection Agency (EPA) to conduct an ecological risk assessment of UV filters to characterize the possible risks to aquatic ecosystems and the species that live in them. However, the 2022 report also identified a number of knowledge gaps and research barriers that may limit understanding of those ecological risks.

To share progress and identify opportunities to further address gaps and barriers, NASEM hosted a workshop in Washington, D.C., on January 23–24, 2023, entitled *Workshop to Advance Research on Understanding Environmental Effects of UV Filters in Sunscreens*. The workshop brought members of government, academia, and industry together in person and virtually to discuss the knowledge gaps identified in the 2022 consensus study, consider research needs specific to the analytical challenges of working with UV filters, understand the environmental effects on aquatic ecosystems and nonstandard organisms, and share possible approaches to standardize toxicity testing.

Through a series of prepared talks, panel discussions, and structured breakout discussions, participants examined the 2022 report and its management context; explored data needs and analytical challenges relevant to the development of accurate toxicity metrics for UV filters; and suggested opportunities to improve and standardize toxicity testing for these chemicals.

These proceedings has been prepared by the workshop rapporteur as a factual summary of what occurred at the workshop. The planning committee's role was limited to planning and convening the workshop. The views contained in the proceedings are those of individual workshop participants and do not necessarily represent the views of all workshop participants, the planning committee, or the National Academies of Sciences, Engineering, and Medicine.

WORKSHOP CONTEXT

Charles Menzie (Exponent, Inc.), chair of the committee that produced *Review of Fate, Exposure, and Effects of Sunscreens in Aquatic Environments and Implications for Sunscreen Usage and Human Health*, set the stage for the workshop with an overview of the 2022 report's key findings and recommendations. Overall, Menzie stressed the committee's finding that further research is needed to better understand the environmental fate of UV filters, reduce uncertainty

---

[1] National Academies of Sciences, Engineering, and Medicine. (2022). *Review of Fate, Exposure, and Effects of Sunscreens in Aquatic Environments and Implications for Sunscreen Usage and Human Health*. Washington, DC: The National Academies Press. https://doi.org/10.17226/26381.

about their potential environmental effects, and enable the higher-tier ecological risk assessments (ERAs) necessary to protect the environments and organisms that interact with them.

The study included a rigorous review of the available data on the processes that influence the fate of UV filters in the environment; measurements in water, soil, and organisms; and environmental effects of the 17 UV filters currently marketed in the United States.[2] The 2022 report emphasized that there is significant variation among these UV filters and, in general, a dearth of knowledge regarding their properties, fate, and impacts. Laboratory studies have found that most UV filters are not biodegradable (with the caveat that studies done in natural water systems may have different results), and high-quality tests show a low-to-moderate potential for bioaccumulation of seven UV filters in aquatic life.[3] However, the wide range of physical and chemical reactions that occur when UV filters enter water, from partitioning to photoreactions to dissolving to settling into sediments, makes toxicity testing and environmental monitoring very challenging.[4] In addition, most of what is known about physico-chemical parameters has been learned from pure water experiments, laboratory conditions, or modeling simulations, and field studies of concentration levels typically lack knowledge of initial exposure in water or sediment. The question of how to perform chemical analyses of UV filters—necessary to understand exposure levels in test systems and the natural environment—was identified as a major knowledge gap in the 2022 report.

Other key issues identified in the 2022 report include the need to better understand exposure and testing dosages, to track degradates and downstream effects, and to develop methods to understand or mitigate the potential for contamination from laboratory testing equipment. The committee also identified a need for methods or benchmarks for measuring acute and chronic toxicity, for better understanding bioaccumulation levels, for characterizing species sensitivity distributions for acute and chronic exposures, and for conducting toxicity tests for nonstandard marine and benthic organisms. The committee also stressed that UV filters do not enter the environment in isolation but within chemical mixtures, pointing to a need to understand the interactions between different chemical components and elucidate the role of coatings, such as silica or aluminum, that are often used with UV filters. Finally, Menzie noted that understanding the impact of UV filters on ecosystems will need to account for the role of other aquatic environmental stressors such as climate change, with particular attention to potential impacts on threatened and endangered species and complex interconnected ecosystems like coral reefs.

The 2022 report included two main recommendations. First, it urged the EPA to conduct ERAs for all currently marketed UV filters and any new ones that become available. The 2022 report outlines critical considerations for these ERAs and noted that the results should be shared with the U.S. Food and Drug Administration (FDA) so that the FDA may include consideration of the environment in its oversight of UV filters in the context of consumer products.

Second, the 2022 report recommended that the EPA, partner organizations, sunscreen formulators, and UV filter manufacturers should conduct, fund, support, and share research and data on sources, fates, environmental effects, bioaccumulation, modes of action, and ecological and toxicity testing of UV filters, alone and in sunscreen formulations. In addition, the 2022 report

---

[2] Organic UV filters include aminobenzoic acid, avobenzone, cinoxate, dioxybenzone, ecamsule, ensulizole, homosalate, meradimate, octinoxate, octisalate, octocrylene, oxybenzone, Padimate O, sulisobenzone, and trolamine salicylate; inorganic UV filters include titanium dioxide (TiO_2) and zinc oxide (ZnO).

[3] Arnot, J. A., & Gobas, F. A. P. C. (2006). A review of bioconcentration factor (BCF) and bioaccumulation factor (BAF) assessments for organic chemicals in aquatic organisms. *Environmental Reviews* 14, 257–297. http://dx.doi.org/10.1139/A06-005.

[4] Schwarzenbach, R. P., Gschwend, P. M., & Imboden, D. M. (2016). *Environmental Organic Chemistry*. John Wiley & Sons.

committee noted that epidemiological risk modeling and behavioral studies are needed to better understand human health outcomes from changing sunscreen availability and usage.

The committee also offered recommendations regarding guidelines for laboratory saltwater experiments; studies of critical body residues for acute and chronic exposure, bioaccumulation, and toxicity; new approaches and methodologies; and robust chemical analytical procedures, such as minimum replicates, standardized collection, extraction, and processing procedures, and quality assurance and quality control measures for both field and laboratory testing to accurately measure UV filter concentrations over time and space.

MANAGEMENT CONTEXT

In its recommendations, the 2022 report committee emphasized the importance of research into the properties and potential environmental impacts of UV filters in order to inform thorough ERAs by the EPA. In addition, many other federal and state agencies may use such data and the resulting ERAs to consider water quality management within a larger context. Gerry Davis (National Oceanic and Atmospheric Administration [NOAA] Fisheries), discussed the environmental management context for ERAs of UV filters and the importance of water quality in general.

Davis's office manages NOAA Fisheries' Pacific Islands Region, which includes the largest, most diverse, and most dense coral reefs within the United States' jurisdiction. NOAA Fisheries also manages four large Marine National Monuments, protected areas comprising some of the last pristine ocean ecosystems and sustaining many endemic and endangered species. Davis stressed that maintaining water quality is essential to the health of both humans and ecosystems and noted that marine systems such as coral reefs are particularly important resources for the global food supply and economy through their impact on fisheries.

NOAA's role in supporting water quality and ecosystem health stems from three main federal laws: the Magnuson–Stevens Fishery Conservation and Management Act, the Fish and Wildlife Coordination Act, and the Clean Water Act. These laws provide the framework within which NOAA weighs the protection of natural resources like coral reefs in the process of considering permits for activities that might affect them. Many other agencies, such as the Army Corps of Engineers and the U.S. Fish and Wildlife Service, also have meaningful roles in managing water quality across the United States and therefore share NOAA's interest in a collective effort to establish adequate, accurate, science-based guidelines for UV filters in sunscreen. Since, as he put it, "you cannot manage what you do not know," Davis said there is great value in advancing research to improve knowledge about the impact of UV filters on water quality and enable ERAs for UV filters that can be used to inform policy development, implementation, and evaluation.

Davis highlighted some specific concerns about the potential impacts of UV filters on coral reefs. Reduced water quality in general, and the addition of UV filters in particular, can disrupt several life stages of the animals that comprise coral reefs. Corals are sedentary organisms and therefore vulnerable to contamination from environmental spikes in water quality and accumulation over time at a fixed site. To manage the impact of sunscreens and other products containing UV filters on corals, Davis said there is a need for water quality standards that are based on knowledge of contamination levels and information about tolerability thresholds, with appropriate consideration for when samples are taken; how UV filters synergize with other contaminants; how lethal or sublethal thresholds are defined; and how factors such as rainfall, wind, and waves affect water dynamics.

Managing the potential impacts of UV filters is further complicated by an overall dearth of knowledge about factors that affect the health of marine ecosystems, uncertainty in understanding water quality, and the influence of other environmental stressors that are simultaneously affecting these ecosystems. For example, discharges of freshwater effluent into oceans can dramatically change salinity, increase volume and velocity, and transport contaminants, which can not only kill coral, prevent fertilization, or disrupt settlement but also interact with UV filters and potentially lead to synergistic effects. Climate change and associated sea level rise and ocean acidification also further complicate the overall environment in which coral reefs are being affected by UV filters and other chemicals.

Noting that researchers and regulators have succeeded in improving thresholds for other water quality stressors such as herbicides and pesticides, Davis posited that there is a need for similar thresholds for UV filters. Such thresholds will need to be based on formal, well-designed ERAs addressing sublethal effects and threshold values, individually and in combination with other pollutants, Davis said, and to be effective they will need to be supported by meaningful and enforceable standards appropriate bodies can use to measure and manage water quality. He added that strong federal support will be needed to help local governments develop and enforce standards.

New tools have made it easier to evaluate the health of living resources and make adaptive changes, a marked improvement over using mortality alone to measure water quality. However, Davis said that capacity, funding, and policy remain obstacles to effectively monitoring and managing water quality. He suggested that greater coordination between federal action agencies and NOAA is needed to overcome these obstacles and identify reasonable, affordable alternatives that keep UV filters out of the water, such as UV-protective clothing. Davis said that managing product availability is likely easier than removing UV filters from water, although he noted that such an approach may be vulnerable to industry litigation.

Finally, although sunscreens have garnered national attention, Davis stressed that water quality is about much more than one product. He urged a broader focus on encouraging collaboration and coordination across federal, tribal, state, and local organizations to more fully address the bigger picture and encourage projects that improve climate resilience and coastal water quality.

UV Filter Chemistry for Accurate Dose–Response Relationships

The workshop's first day focused on the challenges of analyzing and quantifying the structural changes UV filters undergo as they disperse into the environment, along with opportunities to improve understanding of the environmental fate, or final product or effect of these chemicals in an environment. Two speakers set the stage with perspectives on current scientific findings and methods; these speakers were then joined by three additional panelists for a focused discussion examining areas where progress has been made and opportunities for further improvement. Workshop participants then explored these issues further in small breakout groups.

THE ENVIRONMENTAL FATE OF UV FILTERS

Silvia Díaz-Cruz (Institute of Environmental Assessment and Water Research, Spanish National Research Council) reviewed current knowledge about how UV filters are released and their environmental fate. She said that many studies have confirmed the direct release of organic and inorganic UV filters from surface-water contact activities, with increased concentrations correlating with increased activity (such as more people swimming in the summer). UV filters can also be released indirectly from stormwater runoff, industrial runoff, or wastewater treatment plants. The data are extremely limited for all of these mechanisms, however, and there is even less known about potential contributions resulting from illegal dumping or discharge. An added complexity is that UV filters can occur naturally in water and sediment, and are found in many consumer products besides sunscreens, making it challenging to definitively determine sunscreen's contribution to water contamination.

The environmental fates of UV filters vary depending on the physico-chemical properties of the particular UV filter involved, such as its water solubility, volatility, dissociation, hydrophobia, and adsorption, as well as the hydrography and water dynamics of the area in which it is released. Organic UV filters are generally hydrophobic; instead of dissolving in water they partition to particles and sediments. Inorganic UV filters are more likely to aggregate in water and land in sediments, a process that happens faster as water salinity increases.

There is limited information on how UV filters' fates affect the environment. Impacts likely vary depending on environmental parameters such as air temperature, water temperature, salinity, and pH level, as well as how the UV filter interacts with other molecules, its particle size, and its potential for aggregation (see Figure 1).[1] UV filters can disrupt a variety of physical and chemical processes in an environment, such as through direct or indirect phototransformation, or produce potentially toxic reactive oxygen species such as oxygen peroxide. UV filters can also affect biological processes through bioaccumulation in prey, biomagnification in predators, or biotransformations in aquatic microorganisms.[2,3]

---

[1] Schwarzenbach, R. P., Escher, B. I., Fenner, K., Hofstetter, T. B., Johnson, C. A., von Gunten, U., & Wehrli, B. (2006). The challenge of micropollutants in aquatic systems. *Science* (New York, N.Y.), *313*(5790), 1072–1077. https://doi.org/10.1126/science.1127291.

[2] Bar-On, Y. M., Phillips, R., & Milo, R. (2018). The biomass distribution on Earth. *Proceedings of the National Academy of Sciences of the United States of America, 115*(25), 6506–6511. https://doi.org/10.1073/pnas.1711842115.

[3] National Academies of Sciences, Engineering, and Medicine. 2022. *Review of Fate, Exposure, and Effects of Sunscreens in Aquatic Environments and Implications for Sunscreen Usage and Human Health.* Washington, DC: The National Academies Press. https://doi.org/10.17226/26381.

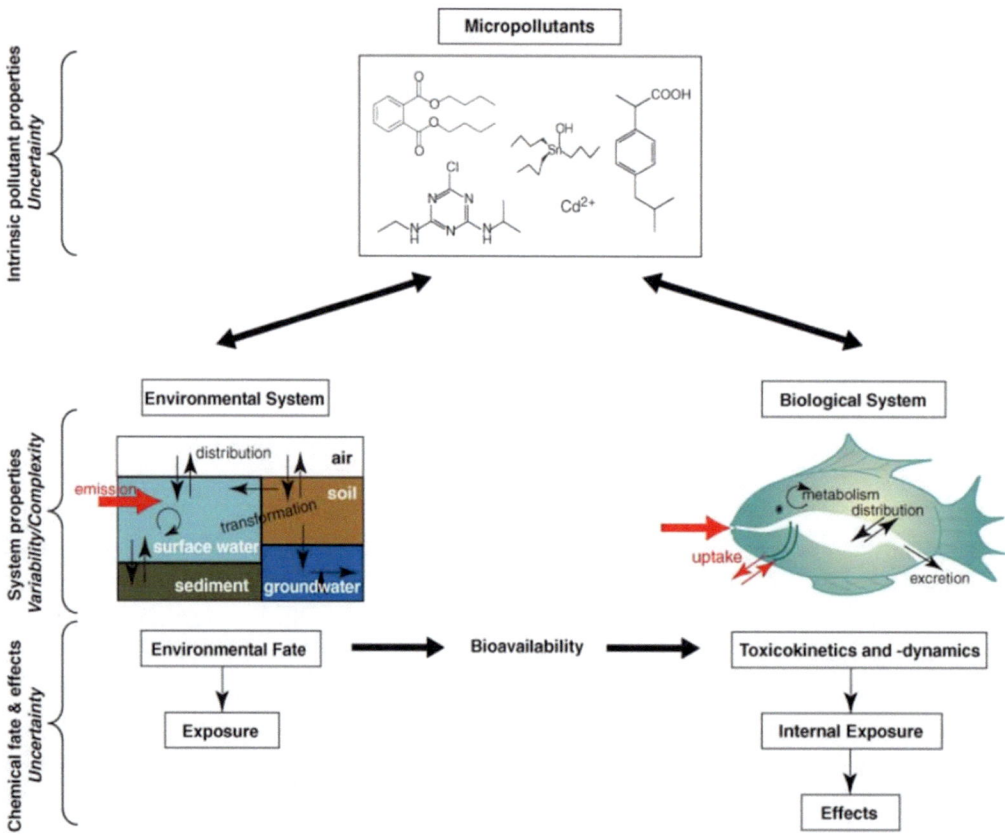

FIGURE 1 Diagram showing the potential fates of common UV filters in aquatic systems. Source: From René P. Schwarzenbach et al., The Challenge of Micropollutants in Aquatic Systems. *Science* 313,1072-1077(2006). doi:10.1126/science.1127291. Reprinted with permission from AAAS.

Given these unknowns, Díaz-Cruz emphasized the need for further research. In particular, she said it is critical to study the potential toxicity of these chemicals, including their metabolites and other transformation products; their biodegradability; the potential for long-chain biomagnification and bioaccumulation; and their impacts on aquatic plants and on critical body burdens to understand long-term biota stress. She also pointed to a particular need for studies to address the impact of inorganic UV filters on soil and aquatic organisms and to elucidate how UV filter coatings, such as aluminum, silica, and polydimethylsiloxane aggregate or dissolve.

Noting that laboratories cannot accurately recreate environmental conditions, Díaz-Cruz said it will be important to conduct field studies to address these questions, including consideration of the specific water dynamics of particular areas. To advance this work, she said there is also a need for researchers to develop and share standardized methods, field blanks, quality controls, and reliable laboratory-based or experimental bioconcentration factors and bioaccumulation factors. In addition, she suggested that monitoring programs should include repeated, replicable measurements over space and time for accurate occurrence data.

ANALYTICAL CHALLENGES IN QUANTIFYING ORGANIC UV FILTERS

Michael Gonsior (University of Maryland) discussed considerations regarding the extraction and quantification of organic UV filters and described particular challenges related to the tautomeric behaviors (when chemical compounds exist in various confirmations simultaneously in a mixture), instability, and solubility of these chemicals (see Figure 2). He noted that these factors impact not only toxicity but also identification; researchers seeking to study a particular compound must consider the possibility that it may have changed due to reaction or degradation, confounding attempts to understand its true fate.

Organic UV filters can transform into tautomers that have very different behaviors in different environments. In addition, UV filters are not stable and tend to degrade over time. Gonsior said that laboratory storage standards are needed to avoid degradation and account for interaction, photoreaction, pH levels, and transesterification, a process in which interacting compounds exchange parts.[4] Quantifying the solubility of UV filters is critical to understanding their environmental fate, he said, but little solubility data are available, and solubility can also vary depending on the environment. In addition, UV filter particles can create a microlayer on the sea surface or other hydrophilic photoproducts instead of dissolving.

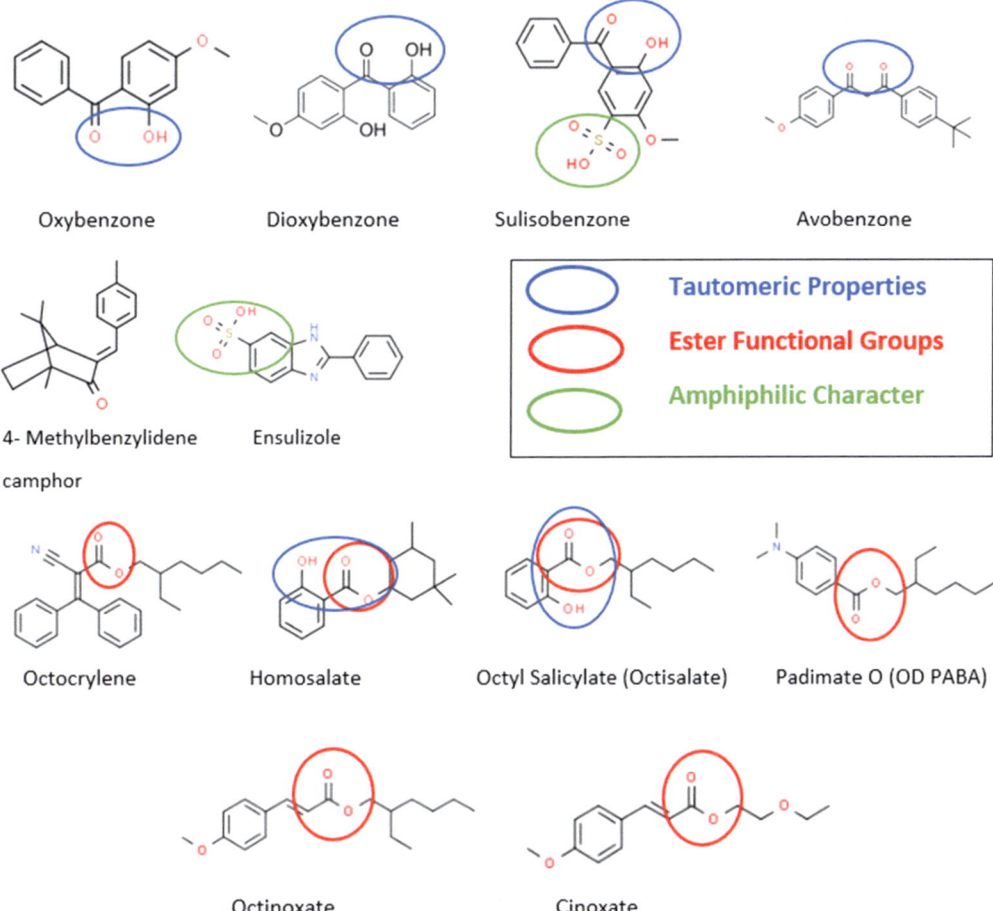

FIGURE 2 Diagrams of common UV filter chemicals highlighting what are tautomeric properties, ester functional groups, and amphiphilic characters. Source: Michael Gonsior's Presentation on January 23, 2023.

---

[4] Holt, E. L., Krokidi, K. M., Turner, M. A. P., Mishra, P., Zwier, T. S., Rodrigues, N. d. N., & Stavros, V. G. (2020). Insights into the photoprotection mechanism of the UV filter homosalate. *Physical Chemistry Chemical Physics, 22*(27): 15509–15519. https://doi.org/10.1039/D0CP02610G.

Extracting UV filters from environmental samples is a key challenge, and Gonsior pointed to a need for extraction standards covering parameters such as whether samples should be filtered. In particular, the use of different isolation techniques, the process of surface adsorption, and contamination from insufficient cleaning or drying can affect results and undermine the ability to compare findings across studies. He suggested that researchers should use EPA recovery standards to report adsorption loss, which some compounds are more prone to than others. In addition, he said that adopting techniques developed to analyze per- and polyfluoroalkyl substance (PFAS) could help to minimize contamination. Gonsior also described a solid-phase extraction method his team developed and applied to measuring 12 common UV filters using chromatography with optimized electrospray ionization. He said the method minimizes adsorption and contamination, avoids false positives, and enables quantification of most organic UV filters at nanomolar detection limits, including curated isotope standards. For future work, he said the team's next priority is to incorporate recovery standards.

PANEL DISCUSSION

Scott Belanger (Procter & Gamble, retired), brought Díaz-Cruz and Gonsior together with three additional panelists to discuss opportunities to overcome the challenges in order to improve understanding the environmental fates of UV filters and inform the chemical analyses necessary to advance ERAs for these chemicals. The additional panelists were Jon Arnot (ARC Arnot Research & Consulting), Bill Mitch (Stanford University), and Kurt Reynertson (Johnson & Johnson Consumer Health). Each panelist offered opening remarks regarding areas of progress and key challenges.

Building on her previous comments, Silvia Díaz-Cruz reiterated the need to study transformation products that react to species and affect toxicity. She said it will also be important to utilize instrumentation including high-resolution mass spectrometers, to develop compound libraries or databases to identify what is visible, to conduct non-target screenings and experiments with the environmental site water and its hydrodynamics, and to further investigate biomagnification higher up the food chain.

Michael Gonsior said that progress has been made in quantifying and analyzing UV filters. To move forward, he reiterated the need to standardize measurement approaches, incorporate EPA recovery standards for different compounds, and conduct lab-comparison studies to produce consistent, reliable, and reproducible data for evaluating toxicity and environmental risk. He also highlighted the value of viewing nontarget screenings across multiple matrices such as freshwater vs. seawater or microlayer vs. dissolved, modeling for toxicity, improving understanding of how filters dissolve or partition, establishing best practices for extraction, and developing better approaches to measure biomagnification.

Jon Arnot highlighted the need for more research into the partitioning of ionizable organic chemicals to solid phases in aquatic environments, including better models of adsorption coefficients and partitioning behavior. He also noted that a better understanding of all of the chemistry involved, such as the implications of tautomerization, is needed in order to extrapolate testing results to environmental effects. Finally, he said that it is important to further study biomagnification in coral, plankton, and other lower-trophic species, though he noted that progress has been made on understanding biotransformations, which will shed light on biomagnification in the food chain.

Bill Mitch stressed the need for new approaches to measure exposure, toxicity, and aqueous concentrations of UV filters and to identify the primary exposure pathways. He said that more research is also needed to understand phase 2 metabolites, create phase 2 standards, and standardize toxicity tests, especially with regard to lighting conditions. He noted that passive sampling has progressed to capture UV filter effects in the water column, provide integrative exposure estimates, and enable measurements of phase 1 metabolites and biomass concentration, which could create a measurable steady-state concentration in tissue.

From the perspective of a sunscreen manufacturer, Kurt Reynertson said that there has been progress in analytical chemistry and contamination identification, but he noted that formulating effective sunscreens that stick to the skin remains a key focus and challenge for product developers. The formulations are complex mixtures of ingredients that can affect UV filters' route into the environment, their fate, and their uptake by organisms. While understanding the intrinsic qualities of discrete elements like UV filters is important, Reynertson posited that it is even more important to understand the mechanisms and potential matrix effects of these complex formulations.

Following panelists' opening remarks, Belanger moderated a discussion addressing testing challenges, how UV filters compare to other chemicals, and opportunities to advance research.

Testing Challenges

Testing UV filters in laboratory settings has multiple challenges that could be overcome so that researchers can, in Belanger's words, "apply the right tool for the right question at the right level of specificity and at the right level of detection." When asked how to minimize sample loss during laboratory testing, Gonsior replied that while some UV filters are more prone to loss than others, if the loss can be identified and quantified, it can be mitigated. Reynertson agreed, noting that it is important to know which compounds are susceptible to loss. Díaz-Cruz added that using weighty standards, in addition to internal standards, during testing enables both loss descriptions and recovery experiments. Mitch noted that for toxicity tests, even test tubes can induce loss, making the critical point—what amount will enter an organism—difficult to measure. He suggested that researchers should use different equipment or passive sampling. Díaz-Cruz agreed but noted that for UV filter analysis, concentrations cannot be interpreted from passive sampling unless the sampling has been validated and at least three or four samples are taken, to account for water dynamics.

Belanger asked if formulation testing of sunscreen would be more valuable than single-chemical testing, despite the analytical challenges. Reynertson said it would, adding that it is these compounds' combination effects that make it especially hard to pinpoint effect sources and availability to organisms. Formulation testing methods differ by company but typically start with single ingredients and then gradually increase complexity. Belanger noted that understanding toxicity is harder when formulas have multiple components and therefore multiple modes of action. In addition, he said, passive dosing methods are not adequately standardized to enable appropriate risk assessment.

How UV Filters Compare to Other Chemicals

UV filters are among many human-made chemicals that can be directly or indirectly transferred into the environment. Mitch noted that his team studies wastewater effluent to determine what compounds are present and in what quantities, but these studies lack predetermined toxicity levels for sunscreens. He and Díaz-Cruz agreed that contaminant quantities are less important than concentration and toxicity levels. Arnot and Belanger noted that while quantification and exposure-route mapping are relevant, it is important to bring the focus back to environmental effects testing for UV filters. For example, UV filters can create a surface microlayer, where Gonsior and his team have found a dramatic concentration factor that can vary depending on wave action. They also found that it is possible for this microlayer's hydrophobic, intertwined chemicals to directly expose coral to these chemicals.

Belanger asked how UV filter compounds are different from other ionizable compounds. Gonsior replied that all ionizable compounds are challenging to work with, and he reiterated that suppression and recovery standards are needed. For example, esters can fall apart during ionization, which occurs differently in seawater and freshwater. Arnot stressed the urgent need to expand understanding of ionization in aquatic and biological test systems to overcome the challenges of studying hydrophobic chemicals, an emerging research area that may be applicable in this space. He also suggested passive dosing, in addition to passive sampling and traditional analytical methods, to learn more about solvent effects and solubility from the samples. Belanger agreed that understanding the timeline of equilibrium and degradation would present important empirical evidence. Sascha Pawlowski (BASF) noted that in addition to passive dosing, saturation columns provide the consistent exposure levels needed to test and create single-compound regulations.

Participant Suggestions to Advance Research

Reynertson underscored the need for more research into how a formulation—not just a particular compound—enters an environment. This information is critical to designing studies of true environmental relevance and toxicity because entry mechanisms also affect UV filter partitioning and organism uptake, he said. Mitch agreed and noted that formulation concentrations can be higher than individual components. He also suggested that a new, integrative approach to measuring exposure is needed, as water concentration can vary by time and space.

To create the robust data that are needed for formal ERAs, Gonsior said that studies need to be further standardized, optimized, and made reproducible. Passive and temporal samples would create a more realistic context, and he suggested that toxicology studies should be paired with analytical chemistry to uncover more of the experience of aquatic organisms. Díaz-Cruz agreed that study standards and protocols are needed, including for sample collection, which can be combined with monitoring to identify what is being measured and determine whether its bioavailability is too low to be taken up by organisms. In addition, she noted that filtering samples will impact the exposure contribution and should be avoided.

Belanger noted that another key area for future research is how UV filters transform during collection, and Gonsior agreed, adding that it is also important to determine how quickly such transformations occur. Mitch added that identifying each compound's mechanism of toxicity is another important research area. To guide well-defined experimental test systems and create useful, relevant information about environmental conditions, Arnot suggested seeking a mechanistic understanding of both toxicokinetics and toxicodynamics to understand discrete organic chemicals and extrapolate their exposure to low-level organisms such as coral, which are generally not well understood. Those studies can then be used to measure internal and external concentrations in the test systems.

Asked to discuss test equipment that could be useful for studies in this area, Gonsior replied that he uses CDN Isotopes for readily available deuterated standards, which offers the suppression, retention, time, and confirmation for each individual compound needed to avoid false positives and optimize chromatography. He also noted that for different solvents, nuclear magnetic resonance and mass spectrometry could be used to measure the concentrations of tautomers, although quantification is somewhat difficult due to degradation.

BREAKOUT DISCUSSIONS

Participants divided into small groups for a deeper dive into specific questions about the challenges of working with UV filters, areas of progress, and examples of key needs to help address the remaining gaps. Representatives from each group summarized the outcomes of these discussions, which are combined and summarized in the sections below.

What Are the Main Chemistry Challenges When Working with UV Filters?

To inform ERAs, several participants emphasized it may be vital to further elucidate the physical and chemical properties of UV filters and the changes they undergo once released into the environment. This includes characteristics and processes such as solubility, partitioning, interactions, transformations, and biodegradation, all within the context of water dynamics and environmental conditions such as salinity. In addition, it is important to understand exposure and delivery pathways, uptake routes, and the role of metabolites and micelles. Many factors complicate a full understanding of these processes, including the wide variability in sunscreen-use patterns by region, variation in runoff amount contributed by different delivery methods, and the challenges inherent in studying contaminants in low concentrations.

Many participants pointed to a lack of interlaboratory standardization of testing, sampling methods, or quality assurance/quality control protocols as a crucial limitation to the research community's ability to close knowledge gaps. Some participants noted that standardization is likely important to appropriately interpret and compare results, especially given the complexities of sunscreen formulations and the fact that research is carried out in multiple laboratories. Recognizing that UV filters are a diverse group of chemicals, some participants said that optimized and verified standards for extraction, handling, and analysis may be beneficial. A few participants also suggested that standardized protocols could include best practices matrices, including for the most sensitive taxa, as microbial organisms metabolize or degrade compounds very differently.

In order to be useful for creating standards and informing policy, several participants expressed a desire for results to be comparable, defensible, and reproducible via multiple methods. However, some participants identified a number of impediments to this goal. For example, in sample testing, one challenge is that each compound has different adsorption rates into glass or plastic labware; another is that the compounds or the solvents may have different purity rates, which affects toxicity; a third is the potential for contamination. Testing acute toxicity is difficult because it is hard to extrapolate bioavailability, toxicodynamics, and toxicokinetics between lab-based and environmental monitoring. Testing chronic or episodic exposure or toxicity might be an easier route, perhaps focusing on marine species that are easiest to analyze, though several participants said that studies of both chronic and acute exposure are important.

A few participants also said that testing could include a minimum level of reporting, recoverable steps, an understanding of loss potential, pulse- or steady-state dosing, and a consideration of potential organism–compound interactions. One idea is to take a "subtraction" approach; instead of studying how these chemicals *add* to an environment, researchers could instead remove them and measure any response. Another option is to develop chemical tracers and standards, similar to the work done to support PFAS assessments.

Are Challenges Magnified When Testing Under Certain Conditions?

Some participants emphasized that testing for contamination by UV filters and related chemicals is a challenging and time-consuming undertaking, whether in the laboratory or in the field. These challenges are compounded by the complexity of environmental variables, such as pH and light levels, which cause different matrix effects and confound measurements. In both laboratory and field settings, conditions can make it particularly difficult to reproducibly study solubility, partitioning, interaction, metabolization, transformation, and concentration fluctuations. In addition, measuring sediment exposure, tissue body burdens, and UV absorption rates from different light sources can further complicate studies.

Testing in the water column amplifies the difficulties of extraction, detection, and determining solubility, partitioning, and toxicity concentrations (a key data point for ERAs, along with temporal resolutions and randomized and diel studies). Testing is easier in a laboratory, some participants noted, but the results would be most useful if they were reproducible, quantifiable, representative of environmental systems and variability, and scalable. Then the results could be used to create a "reference state" applicable across different settings, conditions, and laboratories.

What Progress Is Being Made in Addressing These Challenges?

Progress is being made, and several participants suggested the research community could continue to focus on what *is* known in order to help advance assessments. In terms of addressing knowledge gaps, specific areas of progress discussed by participants include findings related to photostability, biodegradation, and metabolite identification and quantification. In terms of research methods, some participants pointed to the use of reference materials, the development of deuterated compounds and standards, the development of methods for testing in natural conditions, and PFAS passive sampling methods as useful steps forward. In addition, analytical chemistry is improving simultaneous assessment and quantification of compounds, which could help to overcome partitioning, concentration, and toxicity measurement challenges, a few participants noted.

Finally, many participants said that different points of view are beginning to converge and that in general the research community may have a clearer focus on identifying information needs and formulating problems to study. As guidelines and more reproducible ERA data are gradually becoming available, some participants noted that there is a growing awareness that this issue may benefit from significant cross-disciplinary collaborations to make further progress.

What Standardizations, Innovations, and/or Other Focused Efforts Are Suggested to Move Forward on Addressing These Challenges?

Some participants considered a number of opportunities to overcome fragmentation in this area of research and begin to address the remaining gaps and barriers. For example, a participant suggested focused, collaborative efforts by analytical chemistry developers and users to create interlaboratory standardized, centralized, shareable, reliable, and reproducible analytical methods and/or best practices for monitoring, sampling, transport, storage, data usage, reporting, and quality assurance/quality control procedures for studies of UV filter contamination. Other participants suggested a synthetic matrix and standard reference materials, such as for organisms, seawater, water depth level, or background concentration levels, as well as interlaboratory comparison studies and high-resolution mass spectrometry image and video tools. Identifying where biological effects happen, identifying or mitigating cross-contamination, and improving understanding of risk were also suggested as important areas of focus.

To avoid duplication of effort and errors, some participants stressed the benefits for open and transparent sharing of data and methods—for example, by including detailed information about study methodology as supplementary information with published studies, as well as by sharing methods and results from unpublished or "failed" studies. Finally, a few participants underscored the importance of collaboration among government (especially EPA), academia, and industry, supported by appropriate policy and funding along with transparent public reporting on research investments and results.

What Are Existing Research Programs, Capabilities, and Infrastructure That Can Contribute to Addressing Gaps in Research on UV Filter Chemistry?

Participants considered roles for a variety of organizations in enabling further progress toward understanding UV filters and their potential environmental impacts. For example, the National Institute of Standards and Technology could help create needed standards, organizations such as EPA and the Organization for Economic Co-operation and Development could create and disseminate guidelines and tools, and EPA, FDA, and other federal agencies could sponsor grants, workshops, and other initiatives to incentivize and support this work.

Rather than working on individual projects in a vacuum, some participants urged the research community to focus on fostering a coordinated, cross-disciplinary scientific community. Organizations could pool their resources to leverage funding and expertise, and groups already working on exposure, risk, and benefits assessments could work to build stronger and more global collaborations, many participants suggested. Models for multidisciplinary, multisector collaborations exist, including in the area of sunscreens where the research community came together to establish standards around sun protection factors (SPFs).

Examples of particular organizations and programs that could be leveraged to help advance collaborations around UV filters identified by participants include Canada's Experimental Lakes Area, EPA's Safe and Sustainable Water Resources and Chemical Safety for Sustainability research programs, the Florida Department of Environmental Protection, the U.S. Geological Survey, PFAS research programs, and the International Collaboration on Cosmetic Safety. Finally, several participants suggested leveraging the resources of industrial equipment manufacturers like SCIEX and Phenomenex.

Standardizing Approaches for Toxicity Testing

The workshop's second day focused on standardizing approaches for toxicity testing to generate reliable, reproducible data to inform EPA's ERAs on UV filters. To set the stage, Sandy Raimondo (EPA) gave a presentation on the importance of standardized toxicological methods and how EPA uses toxicity test data, which was followed by a series of three short talks on methods for coral ecotoxicology. Participants then delved deeper into challenges and opportunities in toxicity testing for UV filters in a panel discussion and structured breakout group discussions.

EPA PERSPECTIVE: THE IMPORTANCE OF STANDARDIZED TOXICOLOGICAL METHODS FOR AQUATIC ORGANISMS

Raimondo, a research ecologist at EPA's Gulf Ecosystem Measurement and Modeling Division, discussed how EPA uses toxicology testing data, the importance of standardized test methods, and how new approach methodologies can help to generate environmentally and ecologically relevant data that are usable for policy making. "This [NASEM 2022] report is going to spark a flurry of research," she said. "We know this, and we want to make sure that the data that are collected provide information that we can use in a consistent, defensible, and reproducible manner."

In conducting ERAs, EPA typically relies upon data generated from replicable, controlled laboratory studies of a single stressor, but there are scant data of this sort for UV filters. To fill this gap, Raimondo said that researchers could adapt widely accepted standardized test methods to generate conclusive, statistically significant concentration response curves for UV filters for both standard and nonstandard aquatic species. They discussed some particular considerations researchers should be aware of if they want their findings to be useful for EPA and other decision-making bodies (example shown in Figure 3). She stressed that test methods could standardize exposure duration, biological response, age, and life stage for each organism tested. It is also critical, she pointed out, to ensure consistency in terms of water temperature, active ingredient level, salinity, pH level, analytical methodologies, and endpoints in order to generate data that will be comparable across species and laboratories and usable for ERAs. Standardized test methods are different for chronic and acute testing, but in either case, she said it is important to avoid open-ended toxicity values and establish species sensitivity distributions as a quantitative metric for understanding effects.

Raimondo noted that standardized data, the "gold standard" for study defensibility, inform ERAs and improve the understanding of how chemicals affect organisms, which organisms or taxa are most at risk, where EPA should prioritize its resources, and what mitigation effects could reduce the environmental impacts. EPA and others can also utilize such data to establish a chemical's *de minimis* environmental impacts, create water quality standards for officials who grant local permits and conduct environmental restoration, and assess the impacts of accidental chemical releases into the environment. Standardized data also reduce variation when comparing effects across different chemicals and species, helping to guide EPA in prioritizing environmental mitigation and management efforts.

Even standardized methods and data have limitations. Since it is not possible or necessarily ethical to test every species that may possibly be affected by a contaminant, efforts to extrapolate effects across species can help to focus testing efforts on species that are easier to work with. It is useful to examine a wide variety of endpoints, but there are also limitations to the interpretation and extrapolation across endpoints. Finally, Raimondo underscored the persistent challenge of translating between laboratory studies, field studies, and the exceedingly complex physical and biological environments of real-world aquatic and marine ecosystems.

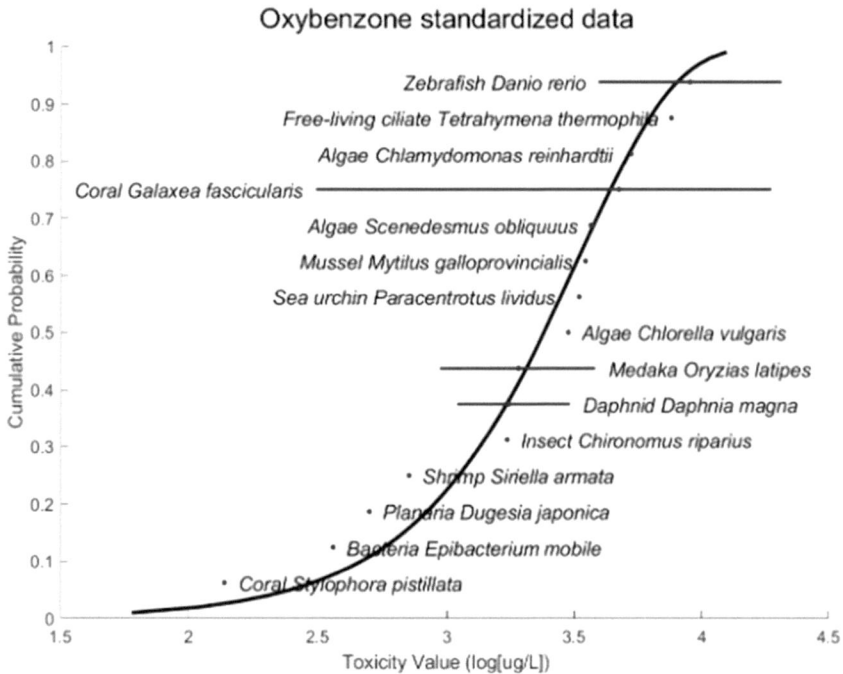

FIGURE 3 Oxybenzone standardized data curve showing where different marine organisms fall on this standardized curve between toxicity value and cumulative probability. Source: Sandy Raimondo's presentation on January 24th, 2023.

In addition to long-established toxicity testing methods, Raimondo said that new approach methodologies can help to obtain reproducible, defensible, and environmentally and ecologically relevant endpoints for UV filter toxicity testing. The most important element of any testing approach, she continued, is to have a consistent, transferable test design with quantitative endpoints that show how organisms respond to a stressor, which can then be used to create higher-tier assessments that improve environmental realism and lead to ecologically relevant conclusions. New approach methodologies can inform toxicity tests that are sensitive to marine environments, account for abiotic variability, and support interspecies extrapolations. In addition, Raimondo said that screening sensitivity assays can lead to more applicable toxicity knowledge in order to prioritize certain chemicals. To move forward, she reiterated the need for interlaboratory consensus on effects and endpoints and stressed the importance of publishing quantitative data and repeating studies.

LIGHTNING TALKS: METHODS FOR CORAL ECOTOXICOLOGY

In a series of three lightning talks, researchers shared specific approaches being explored to refine toxicology studies, particularly in the context of coral.

Considerations for Valid Exposure Designs to Generate Data That Is Relevant for an Era

Craig Downs (Haereticus Environmental Laboratory) discussed four considerations when testing exposure to UV filters for ERA-relevant data: contaminant concentration, exposure medium, exposure timing, and light. Typically, the assumed concentration of UV filters in the water column is measured after a UV filter is added to the water vessel. However, Downs said it is important to recognize that some UV filters move out of the water column to the meniscus layer, vessel walls, or organisms' structural supports and could also be significantly affected by volatilization, all of which can lead to inaccurate measurements of the actual concentration.[1] To combat this problem, he suggested using Teflon liners for beakers and conducting vigorous prescreening tests before introducing an organism in order to understand what confounding leachates may be present in exposure media.

Using the right exposure medium is crucial, Downs emphasized. Noting that filtered seawater is not consistent across sources and often has unknown amounts of xenobiotics, total organic carbon, and other confounding organic matter, he said that a better alternative is for researchers to create their own artificial seawater, for which many established recipes are available. He added that in order to use artificial seawater to approximate the real-world conditions in which UV filters interact with living organisms, it is important to not only approximate the chemical composition of seawater but include known biological and organic components, which themselves affect and are affected by UV filters. Exposure timing also has implications for the handling of exposure media. Noting that tests can be designed with static or semi-static exposures, Downs said that changing the exposure media at set intervals can help to mitigate the issue of loss that can occur over longer exposure periods.

Finally, Downs underscored the important role of light as both a driver of UV filter toxicity and a necessary requirement for culturing and exposure. Standard laboratory lighting does not replicate the natural light organisms experience (light experiment shown in Figure 4).[2,3,4] It is possible to use natural light, though this presents added logistical challenges and protocols necessarily will vary depending on a laboratory's location and environment. Downs suggested moving toward establishing a replicable LED array with justified simulated solar spectrum at model species depth and including true solar incidence angles attenuated with neutral density filters to provide a justified photosynthetic intensity and UV spectrum relevant to the natural environment of the species being studied.

---

[1] Chen, T. H., Wu, Y. T., & Ding, W. H. (2016). UV-filter benzophenone-3 inhibits agonistic behavior in male Siamese fighting fish (*Betta splendens*). *Ecotoxicology* (London, England), *25*(2), 302–309. https://doi.org/10.1007/s10646-015-1588-4.

[2] Collins, P., & Ferguson, J. (1994). Photoallergic contact dermatitis to oxybenzone. *The British Journal of Dermatology*, *131*(1), 124–129. https://doi.org/10.1111/j.1365-2133.1994.tb08469.x.

[3] Downs, C. A., Kramarsky-Winter, E., Segal, R., Fauth, J., Knutson, S., Bronstein, O., Ciner, F. R., Jeger, R., Lichtenfeld, Y., Woodley, C. M., Pennington, P., Cadenas, K., Kushmaro, A., & Loya, Y. (2016). Toxicopathological effects of the sunscreen UV filter, Oxybenzone (Benzophenone-3), on coral planulae and cultured primary cells and its environmental contamination in Hawaii and the U.S. Virgin Islands. *Archives of Environmental Contamination and Toxicology 70*(2), 265–288. https://doi.org/10.1007/s00244-015-0227-7.

[4] Zhong, X., Downs, C. A., Li, Y., Zhang, Z., Li, Y., Liu, B., Gao, H., & Li, Q. (2020). Comparison of toxicological effects of oxybenzone, avobenzone, octocrylene, and octinoxate sunscreen ingredients on cucumber plants (*Cucumis sativus* L.). *The Science of the Total Environment*, *714*, 136879. https://doi.org/10.1016/j.scitotenv.2020.136879.

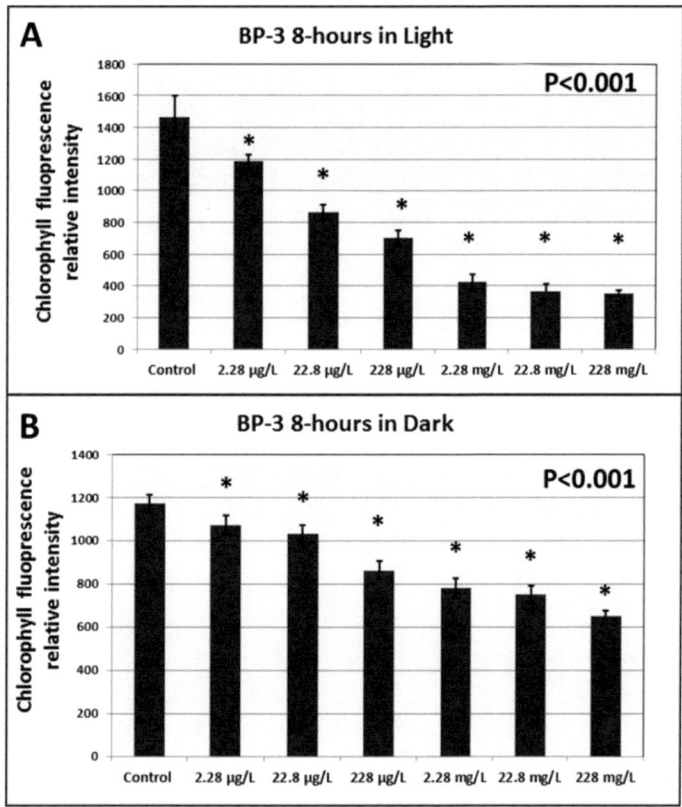

FIGURE 4 Experiment showing light vs. dark reactions. Light acts as both a driver of toxicity and as a necessary requirement for nonconfounding exposure. Source: Downs, C.A., Kramarsky-Winter, E., Segal, R. et al. Toxicopathological Effects of the Sunscreen UV Filter, Oxybenzone (Benzophenone-3), on Coral Planulae and Cultured Primary Cells and Its Environmental Contamination in Hawaii and the U.S. Virgin Islands. Arch Environ Contam Toxicol 70, 265–288 (2016). https://doi.org/10.1007/s00244-015-0227-7, Reproduced with permission from Springer Nature.

Standardization of Toxicity Tests on Corals to Meet Regulatory Requirements

Sascha Pawlowski (BASF) highlighted BASF research relevant to protecting corals from potentially toxic UV filters and discussed how the U.S. context relates to the regulatory context in Europe. He noted that there are nearly 30 UV filters registered in Europe, for which hazard risk assessments are required under the European Union's Registration, Evaluation, Authorization and Restriction of Chemicals (REACH) regulation. Within this framework, if a chemical is deemed hazardous, predicted no-effect concentration (PNEC) levels must be derived, typically using standard laboratory organisms. Until standardized toxicity tests are available, these PNEC levels can be used as surrogates to protect coral, freshwater sediment, and all nonstandard organisms.[5, 6]

---

[5] Pawlowski, S., Moeller, M., Miller, I. B., Kellermann, M. Y., Schupp, P. J., & Petersen-Thiery, M. (2021). UV filters used in sunscreens-A lack in current coral protection? *Integrated Environmental Assessment and Management*, *17*(5), 926–939. https://doi.org/10.1002/ieam.4454.

[6] Pawlowski, S., Lütjens, L. H., Preibisch, A., Acker, S., & Petersen-Thiery, M. (2023, submitted). Cosmetic UV filters in the environment–State of the art in EU regulations, science and possible knowledge gaps. *Journal of Cosmetic Science, Special Series: Sun Protection.*

To work toward standardized toxicity tests for future regulatory requirements, key aspects to be considered are shown in Figure 5. BASF is adapting existing guidelines, including guidelines from the Organisation for Economic Co-operation and Development (OECD) and the International Organization for Standardization (ISO) for new test organisms. In designing these tests, he said it is important to consider all life stages, from larvae to adulthood; to account for both acute and chronic toxicity; to discriminate between intrinsic toxicity and physical effects; and to use population-relevant endpoints such as mortality, metamorphosis, and growth. In terms of methodology, he added that it is important to ensure water quality and light conditions are suitable for coral survival, to avoid unnatural and potentially confounding solvents, to use chemical control analysis to refer the results to nominal concentrations, and to measure the stability and the recovery of the substance at the end of the exposure regime.

In collaboration with the University of Oldenburg, Germany, BASF has conducted short-term acute toxicity tests on lab-bred adult corals and longer tests on more sensitive larvae, which Pawlowski said are ready for prevalidation under ISO/OECD. Researchers have also studied longer-term effects of certain solvents, such as dimethylformamide, dimethyl sulfoxide, ethanol, and methanol. Chronic toxicity tests for coral fragments are in progress, and the team next plans to focus on coral bioaccumulation tests, with the goal of developing a standardized coral toxicity testing method that generates high-quality data in the next 5–10 years, Pawlowski said.

Applied Development of Standardized Coral Toxicity Tests

Abigail Renegar (Nova Southeastern University) provided additional context on the challenges of developing standardized coral toxicity tests. Corals are highly sensitive to factors such as temperature, light, water flow, salinity, pH, and alkalinity and also exhibit a wide range of chemical responses, within species and during certain life stages, all of which makes them quite difficult to study. As a further complication, corals have a symbiotic relationship with zooxanthellae, making it necessary for study endpoints to consider impacts on both organisms, she said.

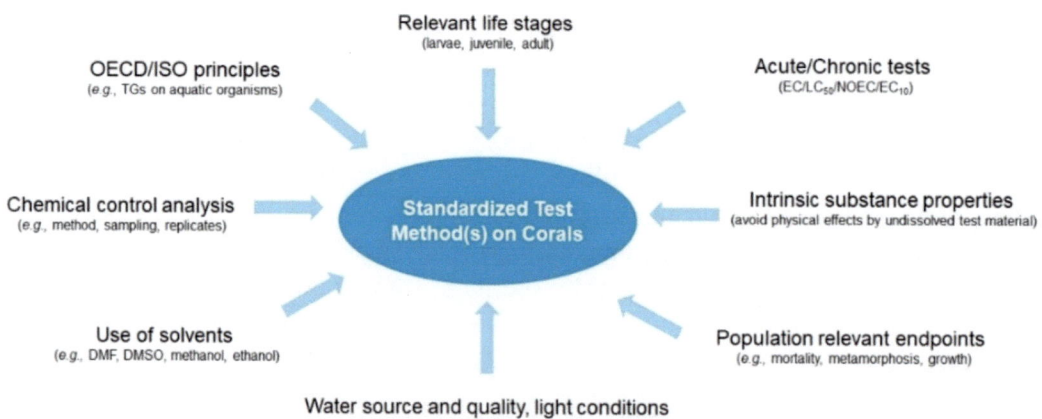

FIGURE 5 Diagram showing the key aspects to be considered for standardized test methods on corals. Source: Sascha Pawlowski's presentation on January 24th, 2023.

Despite the challenges, researchers have generated highly reproducible toxicity thresholds in corals based on histological changes and growth rate. However, Renegar cautioned that generating reproducible, low-variability, high-quality data require significant attention to detail, minimization of coral handling, and further optimization of methods.

Regardless of the particular chemical or coral species being tested, toxicity assessments generally follow the same essential structure. First, pre-exposure observations are used to establish a baseline of growth rate and photosynthetic efficiency. Then, measurements during exposure or immediately post-exposure are used to assess mortality, coral condition, polyp behavior, mucus production, coloration changes, tissue swelling or attenuation, growth rate, and photosynthetic ability. Finally, post-exposure measurements are used to assess bioaccumulation, gene expression, and histological changes such as tissue architecture, cellular integrity, and zooxanthellae condition. Within this overall structure there are a number of testing approaches, each of which has its own set of strengths and weaknesses. For example, Renegar said that static-renewal tests are useful for finding and assessing the range of responses to new chemicals or working with a new species but tend to underestimate toxicity. Continuous recirculating systems are closed, use passive dosing, and are useful for short, acute exposures to volatile and semivolatile compounds. Flow-through methodology is useful for 96-hour acute and 21-day chronic exposures, maintains high water quality, and results in reliable and consistent exposure concentrations for chemicals that are difficult to work with, like UV filters.

To continue to refine and standardize testing approaches, Renegar said that it will be vital to overcome the challenges of analytical chemistry and balance testing standards and consistency with existing regulatory guidelines and metrics that are scalable for environmentally relevant exposures. Atlantic staghorn coral (*Acropora cervicornis*) is one of Renegar's test species to show coral condition endpoints from environmentally relevant exposures of UV filters (Figure 6). She suggested prioritization of corals and other nonstandard organisms of significant ecological importance in the context of UV filters and also suggested that EPA and OECD should develop standard coral toxicity test methods for chronic exposure with population-relevant endpoints. She added that the scientific community as a whole should focus on producing reliable, actionable data with appropriate quality assurance/quality control reporting standards and noted that molecular methods may also be applicable.

FIGURE 6 Pictures of coral condition endpoints of *Acropora cervicornis* (Atlantic staghorn coral) after exposure to UV filters. Source: D. Abigail Renegar's presentation on January 24th, 2023.

PANEL DISCUSSION

Transitioning to a panel discussion, moderator Carys Mitchelmore (University of Maryland) reviewed the challenges of analytical testing of UV filters that were brought up during the workshop's first day and emphasized the need for multidisciplinary, interlaboratory collaboration to make progress. She then invited an additional five panelists to join Raimondo, Downs, Pawlowski, and Renegar for a discussion of progress, opportunities, and outstanding challenges relevant to aquatic toxicology of UV filters. The additional panelists were Iain Davies (Personal Care Products Council), Marc Leonard (L'Oréal), Mandy Annis (U.S. Fish and Wildlife Service), Jeffrey Steevens (U.S. Geological Survey), and Dan Villeneuve (EPA). Together with Mitchelmore, panelists shared opening remarks and then engaged in an open discussion to bring together perspectives from a wide range of organizations and sectors.

The environmental safety program of the Personal Care Products Council relies on regulatory guidance to assess the environmental risks of UV filters. Davies said that more testing is needed to generate relevant, reliable data for many UV filters, especially for understanding chronic exposures and understanding effects on underrepresented taxa. In Europe, he noted that dynamic predicted no-effect concentrations are required for acute and chronic exposure and for the most sensitive and environmentally relevant species. In the United States, EPA's Office of Prevention, Pesticides, and Toxic Substances has helpful risk assessment frameworks and problem formulations for identifying, testing, and generating reliable data on standard and nonstandard species. He added that other frameworks, such as the Criteria for Reporting and Evaluating Ecotoxicity Data and Canada's Ecoquest, could also be helpful to consult.

L'Oréal has developed a variety of methodologies for chronic toxicity tests of chemical compounds. To develop approaches relevant to the highly complex context of coral reefs, Leonard stressed that multiple challenges should be overcome. In particular, he pointed to the need to take into account the nuances of particular species and the influence of environmental factors such as light length, intensity, distribution, and positioning poses special challenges. Leonard also underscored that new test methodologies for coral should be biologically relevant to the taxonomically diverse assemblages of organisms that comprise coral reefs and stressed the importance of using species sensitivity distributions to identify and test the most sensitive species.

Resource managers with the U.S. Fish and Wildlife Service draw upon the best available data to conserve, protect, and enhance fish and wildlife resources and their habitats. Annis highlighted her team's work with freshwater mussels, a species often considered to be the backbone of many freshwater aquatic ecosystems. Mussels, which improve water quality and stabilize cycling streams with their filtering activity, are similar to corals in that they are extremely sensitive to environmental perturbations and have a complex life cycle that makes them challenging to study. To stave off mussel population declines, researchers are working to create a baseline characterization of mussel health across life stages, establish suitable surrogate species, and develop standardized procedures to study lethal and sublethal exposure levels and durations.

Building on these points, Jeffrey Steevens made a case for broadening the focus of toxicity testing for UV filters to better account for the full diversity of species that make up aquatic ecosystems. The taxa that may be affected by these chemicals live in extremely complex ecosystems, with flowing sediment, terrestrial interfaces, and both static and moving water. He pointed to freshwater mussels (which burrow into sediment and filter water at a high rate) and mayflies (which emerge through the surface microlayer in their early life stages) as examples of species with unique considerations in the context of their potential interactions with UV filters and related chemicals. He noted that impacts on such species could potentially be missed with toxicity testing methodologies that are more narrowly focused on the water column.

Villeneuve noted that it will take time and resources to fill the many data gaps with regard to UV filters and their environmental impacts. He said that new approach methodologies can potentially generate data more rapidly and cost effectively, can be combined with modeling to make predictions about other species that cannot be tested, and can potentially generate accessible data for use in ERAs. However, he cautioned that existing new approach methodologies include many uncertainties, especially around lab-to-field extrapolation, making them perhaps best suited to lower-tier assessments. He suggested combining these approaches with more traditional approaches and placing high priority on appropriate problem formulations, key UV filter data gaps, coral sensitivity, and species sensitivity distributions.

Raimondo, Downs, Pawlowski, and Renegar briefly expanded upon their previous remarks to transition into the panel discussion. To develop the most defensible ERAs, Raimondo stressed that EPA needs the best science available, rooted in minimal variability and strong reproducibility, consensus on endpoints, and an emphasis on ecological and environmental realism. Downs built on his remarks about the importance of standardizing the use of light—a major driver of coral growth and UV filter activity—in toxicity testing. He pointed to a need for EPA or OECD standards on light, which he said should be designed to emulate the light composition and intensity test species encounter at different times of day and at different geographical locations.

Pawlowski emphasized that compromises will be necessary to create EPA/OECD testing protocols that create reproducible results, include appropriate endpoints, and achieve consistent test conditions and realistic species sensitivity distributions. Recognizing that most laboratories around the world do not have dedicated expertise in working with corals, he said it is especially important to establish a standardized framework for testing that can be implemented broadly by people with varying levels of expertise and experience. While progress has been made and some standard guidelines exist, Renegar reiterated that standardized EPA and OECD compatible methods for nonstandard species like corals remain a key need. She said that advancing this work will require multiple partners, significant investments of time and resources, a consensus approach to endpoint selection, data reliability requirements, consistent methods for analytical chemistry, chronic assays, and an understanding of mixed or multiple stressors.

Closing out the panelists' introductory remarks, Mitchelmore stressed the critical importance of advancing analytical methods for toxicological evaluation that are appropriate for the test chemical and exposure design and considerate of factors such as UV filter solubility, solvents, and photosensitivity. She said that protocols should include steps to minimize opportunities for mistakes, contamination, mismeasurements, and misinterpretation. These methods can be adapted from existing EPA and OECD standards and guidelines, with modifications for nonstandard species and adequate quality assurance/quality control in order to enable reporting of reliable, relevant data for ERAs.

Mitchelmore moderated an open discussion between panelists and workshop attendees that further explored toxicity testing challenges, the role of new approach methodologies and species sensitivity distributions, the importance of creating standards, and considerations specific to working with corals and mussels.

Testing Challenges

Developing appropriate study designs and exposure regimes for toxicology testing poses significant challenges. Steevens noted that flow-through testing establishes constant-exposure concentrations, which are important for studies of longer duration, although for those studies, researchers should also add food, which influences exposure. Mitchelmore agreed, adding that organisms also excrete fluids, representing another variable, although the water movement in flow-through systems keeps the exposure more homogenous.

Pawlowski highlighted the importance of the materials that are used, as certain plastic vessels could be problematic. Mitchelmore said that flow-through system standards require glass labware to maintain water quality, keep exposure homogenous, and minimize loss and contamination. Downs reiterated that artificial seawater is better than filtered seawater, especially for acute tests, to eliminate the presence of confounding chemicals. Renegar agreed that using artificial seawater eliminates a source of variability, and Pawlowski noted that he is aware of several labs that use artificial seawater to culture corals and for chronic testing.

Renegar noted that creating realistic flow-through tests with optimized dosage rates requires intensive sampling, time to reach equilibrium, and appropriate consideration of loss. Avoiding loss of control organisms is critical, Raimondo said, because nonoptimized conditions can affect toxicity understanding and lab-to-field extrapolation. Mitchelmore agreed and suggested that coral researchers should adapt conditions for standard organisms, where many of these variables are prescribed to control survival.

Another challenge is addressing contributions from multiple stressors. Renegar stated that scientists learn by understanding one thing at a time, so each stressor should be studied individually first, then combined with others in stepwise fashion.

New Approach Methodologies and Species Sensitivity Distributions

Once relevant, reliable data have been generated, next steps for UV filters can be determined, but Davies noted that this process will take time. To speed progress, problem formulation is an important consideration. Since it is not possible to test all chemicals on all organisms every day to create ERAs or hazard assessments, Davies suggested that temporal and spatial exposure assessments and monitoring can provide more data to home in on problem formulations that define which organisms should be tested, where they live, and which data points are needed.

Villeneuve noted that new approach methodologies go beyond the conventional expose-and-observe technique to more predictive approaches, such as studying an organism's molecular biochemistry and interpolating hazards or population-level or ecosystem-level modeling for outcome prediction. Raimondo reiterated that EPA is working to develop new approach methodologies to overcome many of the challenges identified in this workshop, both for UV filters and for other chemicals under EPA's purview. These approaches are promising for filling data gaps, especially for understudied species, but she underscored the need for more testing in order to reduce uncertainty to a level appropriate for an ERA. "Any approach is only as good as the data that you put into it," Raimondo stated.

Participants explored the benefits and downsides of focusing on species sensitivity distributions. Steevens pointed out that one inherent flaw of this approach is that, by definition, it aims to protect 95 percent of species and requires readjustment if more sensitive species are identified, as has happened for mussels. Many difficult-to-culture species are never tested, and without that data the species sensitivity distribution may underestimate the effects. Raimondo added that the wide taxonomic diversity within aquatic ecosystems makes it difficult for species sensitivity distributions to account for what is unknown, but she said this is an area in which new approach methodologies may help to fill gaps. Pawlowski stated that while species sensitivity distribution data are widely available and commonly used in Europe to protect nontested species, the huge range of variability can reduce confidence in species sensitivity distribution assumptions. Davies suggested European and U.S. safety assessment factors could be merged by replacing organisms with no observed effect concentration (NOEC) with organisms with 10 percent effect concentration (EC10) to improve the species sensitivity distribution and reduce uncertainty.

Creating Standards

To work toward standardization in toxicity testing for UV filters, Raimondo stated that EPA is focused on reducing sources of variability or uncertainty in terms of factors like pH level, light, salinity, test chambers, water quality, and temperature, all of which can affect analyses. Mitchelmore agreed that water quality standards should include baseline exposure media to keep water quality stable. Pawlowski said that reducing variability provides data on intrinsic control, but he cautioned that in many cases ecological relevance is still uncertain and data on corals, especially for longer exposure periods, is still lacking. He added that creating universal toxicity test standards will require collaboration and guidance from relevant authorities.

Raimondo said organizations like EPA and OECD look to "round-robin" testing, in which multiple institutions apply the same methods and get similar results, to ensure standards are transferable and reproducible. However, Raimondo and Mitchelmore noted that the time required for this "gold standard" approach poses a challenge as it can take 5–20 years to fully refine and standardize methodology for existing and new taxa. While Raimondo said that it is important to continue to share protocols and lay the groundwork for round-robin testing in the long run, Mitchelmore suggested that convening a working group to develop informal standard protocols could be a helpful near-term step to establish a set of minimum standards and facilitate comparison across studies. Pawlowski and Mitchelmore added repeating studies and publishing the methods and results in detail is critical to increasing transparency and working toward standard protocols.

Special Considerations for Corals and Mussels

Participants discussed several particular considerations for working with corals and mussels. Whether a substance is toxic or not depends on how toxicity is defined. Asked whether settlement should be considered a key part of coral's life cycle, Pawlowski replied that for short-term mortality studies, settlement is a key endpoint that is probably affected at any concentration. Renegar reiterated that corals are complex organisms and suggested that the microbiome that develops around them should be studied extensively to understand the cumulative effects on the whole ecosystem. For corals, polyp retraction is a common response to contaminant exposure, but its ecological relevance and long-term health impacts are unclear. Mitchelmore agreed that such responses could either be an early sensitivity indicator or a change that does not impact health, and Pawlowski noted that every substance that enters an organism causes a physiological response. Davies suggested that the debate over appropriate toxicological endpoints for corals underscores the need for greater collaboration.[7]

---

[7] Burns, E. E., & Davies, I. A. (2021). Coral ecotoxicological data evaluation for the environmental safety assessment of Ultraviolet Filters. *Environmental Toxicology and Chemistry*, *40*(12), 3441–3464. https://doi.org/10.1002/etc.5229.

Asked about current progress in toxicity testing of UV filters in mussels, Steevens said that researchers have identified a promising test species and determined sensitive endpoints. The next step, he said, is to determine how to prepare test solutions, observe how the mussels respond, and perform bioassays to establish standard thresholds. Annis added that because mussels are challenging to study, much of the available data reflect single snapshots in time. It is important to continue to learn about multiple life stages through chronic assessments to determine biologically relevant, longer-term exposure effects from chemical stressors, she said.

BREAKOUT DISCUSSIONS

In the workshop's final session, participants divided into small groups for focused discussions around the key challenges of working with nonstandard organisms and endpoints and opportunities to make progress in addressing these challenges. Representatives from each group summarized the outcomes of these discussions, which are combined and summarized in the sections below.

What Are the Main Challenges Encountered When Working with Nonstandard Organisms or Endpoints?

Several participants pointed out that the standard species used in toxicity testing were selected as standards based on a number of factors, including the ease with which they can be accessed and used. The challenges of identifying nonstandard species to target, reliably accessing those organisms, and resolving practical considerations for culturing them in a laboratory are not trivial. For many nonstandard species, the lack of a baseline understanding of behavior, genotype, characteristics, life cycles, and sensitivity hampers the ability to design culture conditions that mimic the natural environment and conduct toxicity tests that are appropriate and relevant. It is also important to consider complex interactions such as symbiosis, the role of the microbiome, and other indirect effects on organisms' biology and community interactions, as well as the potential for seasonal variability, both in an organism's natural environment and in potential UV filter exposure patterns, some participants noted.

In selecting endpoints, a few participants suggested that researchers would benefit from defining which responses are considered to be adverse, what sublethal endpoints should be considered, and how laboratory measurements relate to population, growth, or other indicators of viability. The timescale of both exposures and effects is also important. Many participants noted that studying modes of action, while not always essential to toxicity testing, can help to inform the design of future UV filters to avoid causing toxicity via the same mechanisms.

Additional challenges may include developing standardized new approach methodologies, eliminating technician bias, ensuring results are comparable across experiments and laboratories, establishing minimum reporting standards, and developing and validating methods for extrapolating for species that remain difficult to study.

Are Challenges Magnified When Testing under Certain Conditions for Both Standard and Nonstandard Tests for These Chemicals?

Several participants said that these challenges are magnified by a dearth of baseline knowledge about UV filters, the ecosystems that they interact with, and the best ways to develop laboratory tests with ecological relevance. The limited understanding of the unique chemical properties of UV filters, how they interact with seawater, and how they act in mixtures makes it particularly difficult to design tests that will extrapolate to real-world environments. Some participants pointed to for the importance of more research to elucidate the dynamics of partitioning, non-monotonic chemical responses, and solvents' fates and interactions with organisms. Finally, a better understanding of the effects of timing (and related considerations, such as moon phases) and how the medium affects exposure can help to inform study design and reduce the likelihood of loss or contamination.

What Progress Is Being Made in Addressing These Challenges?

Some participants said that the very existence of the 2022 report and this workshop represent important steps toward creating more reliable and reproducible data collection and analysis standards. In addition, several participants said that toxicokinetic and toxicodynamic modeling approaches have improved opportunities to estimate internal tissue exposures and suggested that the tiered modeling exposure frameworks from the International Cooperation of Cosmetic Safety (ICCS) can help to translate direct releases into exposures. The Society of Environmental Toxicology and Chemistry's (SETAC) aquatic testing subgroups, ICCS' coral working group, the Sanger Institute's coral genome sequencing studies, and others have also made important strides in the field, some participants noted.

A few participants also said the development of and increased access to mathematical tools and technologies as well as diagnostic multi-omics literature have helped address some of these challenges, along with improvements in fact-finding testing to guide nonstandard methods, improved coral husbandry techniques, and data sharing.

What Standardizations, Innovations, and/or Other Focused Efforts Are Needed to Move Forward on Addressing These Challenges?

Many participants pointed to the importance of a centralized, global, appropriately funded interagency collaboration involving industry, regulators, and academic researchers to determine what test methods are available, identify the most defensible, and avoid repeating mistakes. Several participants discussed roles for a variety of stakeholders, suggesting that a champion may be needed to help propel a broad collaboration; working groups could be useful for creating action steps and disseminating findings; sponsors could help to support new facilities, produce standardized specimens, and establish research incentives, scholarships, and internships; and data repositories could help increase data accessibility with proper attribution, ownership, and reliability protocols.

From a scientific perspective, some participants underscored the importance of improving problem formulation. For this, it would likely be helpful to identify areas where exposure is highest, discover other exposure pathways beyond sunscreens, and generate robust, statistical endpoint data that would be useful in EPA ERAs. In terms of specific suggestions for future research, some participants mentioned investing in more studies of benthic organisms, sediment, and freshwater species; refining methods to culture nonstandard organisms for laboratory testing; studying multiple stressors separately; and working to further enhance, validate, and increase the use of new approach methodologies. A few participants also suggested expanding into *in vitro* assays and ground truthing, experimenting with cryopreservation, generating toxicokinetic evidence, updating water resistance testing, and standardizing animal husbandry techniques. Pilot studies and validation studies can help to clarify endpoints, refine test protocols, and ensure reliability. Finally, many participants pointed to the value of including absorption, distribution, metabolism, elimination, and reference compounds in study designs and reiterated the importance of carefully considering species selection to ensure species are geographically appropriate and including consideration of endangered species.

To move toward standardized tests and inform ERAs, a few participants suggested learning from biologists, especially coral biologists, but also suggested looking to other countries' standards, studies of toxicology, existing tools, and exposure models in other domains. Some participants suggested broadening ERAs to incorporate a wider range of evidence types and many participants reiterated the desire to find the right exposure media; identify the most important variables; use mesocosms to enable dose dependent responses, transportation, and location in the water column; use field exposures to give environmental context to laboratory tests; estimate biological responses for nonstandard organisms via modeling and nonstandard metrics for sublethal effects; and ask experts to translate toxicology work into real-world scenarios.

From a broader perspective, some participants mentioned the usefulness of deep reflection on industry responsibility and ethics, a better understanding of consumer behavior, a holistic approach to ecosystem stability, an acknowledgment that animals should not be sacrificed unnecessarily, and a push for more open science practices.

What Are Existing Research Programs, Capabilities, and Infrastructure That Can Contribute to Addressing Challenges in Gaps in Research on UV Filter Toxicity?

Many relevant organizations have the resources and expertise to advance collaborative efforts in this space. In particular, a few participants pointed to EPA, OECD, FDA, SETAC, ICCS, ISO, the U.S. Fish and Wildlife Service, and ASTM International as potential key players. In addition to regulations in other countries, some participants said the field can find useful models in EPA's Science to Achieve Results (STAR) program, the Great Lakes Restoration Initiative (GLRI), the "reef safe" designation, the oil industry's experience with spills, and eco-epidemiology approaches that are used to characterize stressors and determine causes of decline to understand multi-stressor impacts on corals. Finally, many participants emphasized that the public is a stakeholder whose interest can spur action, prioritization, and funding to create a better understanding of the risks associated with UV filters and inform the path forward.

Closing Remarks

The two-day workshop provided a forum for lively multidisciplinary, cross-sector discussions building upon the 2022 report *Review of Fate, Exposure, and Effects of Sunscreens in Aquatic Environments and Implications for Sunscreen Usage and Human Health*. Participants explored the current state of knowledge and future research areas to gather data to inform EPA ERAs of UV filters.

Reflecting on the workshop's first day, Menzie said that the discussions of analytical methods crystallized the many difficulties of understanding and measuring the environmental effects of UV filters. Participants delved into the challenges of establishing the sampling and analytical methods to characterize real-world exposures. This includes addressing aspects of how UV filters enter water and in what magnitudes, and how that varies over time and space. It also includes attention to sunscreen formulations, mixing in the environment, and how the interactions between chemicals might influence the fates of different UV filters. Participants also discussed the complexities of understanding routes of exposure in water and sediments, how relevant chemicals may transform or accumulate in biological systems, and novel and emerging research methods for advancing knowledge in these areas.

The workshop's second day examined standardized approaches for toxicity testing of UV filters. With a particular focus on the complexities of working with corals, presenters, panelists, and other workshop participants discussed a wide range of challenges in toxicological testing involving nonstandard species and nonstandard endpoints. Participants considered recent findings and approaches that can help to address some of these challenges and discussed future research areas that could help to further refine protocols. Participants also discussed broader opportunities to facilitate the sharing of testing methods and data and lay the groundwork for reliable, defensible data that can be used for ERAs.

In their discussions on both days, many participants mentioned the importance of multidisciplinary, cross-sector collaboration to help standardize methods to yield reproducible and reliable results across laboratories and to help create a baseline level of understanding of the effects of chronic and acute exposures to UV filters and related chemicals. Closing the workshop, Mitchelmore emphasized the desire to continue conversations among researchers, industry, and government bodies about how best to fund and conduct the needed research, disseminate findings, and support informed decision-making.

Appendix A
Statement of Task

An ad hoc committee of the National Academies of Sciences, Engineering, and Medicine will review the state of science on use of sunscreen ingredients that are currently marketed in the United States, their fates and effects in aquatic environments (focusing on U.S. aquatic environments but with consideration of international studies), and the potential public health implications associated with reduced use. For this review, UV filters will be considered broadly in terms of active ingredients and formulations.

Section 1: Review of fates and effects in aquatic environments. This section will be organized to provide information for future application in ecological risk assessment as outlined below:

1. Problem Formulation: provide chemical profiles for UV filters and applicable degradates (environmental and metabolic), including mode of action for intended use and indirect mode of action where known.
2. Exposure Analysis: identify sources and relative quantities of UV filters entering the variety of aquatic environments (e.g., estuary, lake, coral reef), their fate and transport, as well as measured concentrations in these environments and the biota; identify potential routes of exposure to UV filters and their degradates and the potential for bioaccumulation in aquatic organisms.
3. Effects Analysis: identify potential effects of UV filters on aquatic organisms, including potential for endocrine disruption, photo-activation, and other reported effects on molecular, cellular, organismal, population, and/or community-level endpoints; identify organisms that are listed (Endangered Species Act) or are the subject of targeted management plans that have a high likelihood of exposure; and assess ecosystem-level impacts of UV filters on ecologically, economically, and commercially important habitats (e.g., coral reefs, eel grass beds).
4. Identification of Research Needs: determine information gaps in the above listed areas to identify research priorities to inform both screening and higher-tier ERAs.

Section 2: Implications of potential changes in sunscreen usage on public health. This section will review and summarize the available literature on the use of sunscreen to prevent skin damage in humans from excess exposure to UV in sunlight, including:

1. Summary of information on chemical and mineral UV filter efficacy in preventing UV damage to humans.
2. Potential for changes in sunscreen usage through reduced or less frequent application of UV filters based on concerns about the possible environmental impacts or ease of use of alternative sunscreen formulations. The report will consider use patterns associated with aquatic activities and for outdoors activities in general.
3. Anticipated health consequences of abstention from or reduced use of currently marketed sunscreen ingredients or substitution of alternative UV filters.

Following release of the committee's report, a workshop will be organized to disseminate and further explore the research needs identified in the report. The presentations and discussions at the workshop will be documented in these workshop proceedings, written by a designated rapporteur in accordance with institutional guidelines.

Appendix B
Workshop Agenda

WORKSHOP TO ADVANCE RESEARCH ON UNDERSTANDING ENVIRONMENTAL EFFECTS OF UV FILTERS IN SUNSCREENS

JANUARY 23-24, 2023
Keck Center of the National Academies Room K100

500 Fifth St. NW
Washington, DC 20001
With virtual participation

Purpose

- Disseminate findings from the 2022 National Academies report, *Review of Fate, Exposure, and Effects of Sunscreens in Aquatic Environments and Implications for Sunscreen Usage and Human Health.*
- Discuss the knowledge gaps identified in the report related to understanding the potential effects of UV filters on aquatic ecosystems.
- Serve as a forum for sharing progress on this topic from the public, private, and academic sectors to fill priority knowledge gaps and identify areas of opportunity for further efforts across all sectors.

PRERECORDED PRESENTATIONS

The following prerecorded presentations will be available in advance of the workshop at:
https://www.nationalacademies.org/event/01-23-2023/workshop-to-advance-research-on-understanding-environmental-effects-of-uv-filters-from-sunscreens

Expected availability January 9, 2023.

Findings and Knowledge Gaps from *Review of Fate, Exposure, and Effects of Sunscreens in Aquatic Environments and Implications for Sunscreen Usage and Human Health*
Charles Menzie, Exponent, Inc., Committee Chair

Information Needs for Environmental Management
Gerry Davis, Pacific Islands Regional Office, National Oceanic and Atmospheric Administration

---

MONDAY, JANUARY 23, 2023

10:00–10:20   **Welcome and Meeting Goals**
                Charles Menzie, Exponent, Inc., Committee Chair

Session 1: UV Filter Chemistry for Accurate Dose–Response Relationships

10:20–10:45   **Environmental Fate of UV Filters**
                Silvia Díaz-Cruz, Institute of Environmental Assessment and Water Research

10:45–11:05	**Analytical Approaches for UV Filters** **Michael Gonsior,** University of Maryland Center for Environmental Science
11:05–12:35	**Panel** *Format: Panelists will provide prepared remarks and then participate in discussion regarding progress, opportunities, and outstanding challenges relevant to UV filter analytical chemistry.* **Moderator: Scott Belanger,** Procter & Gamble (retired), Committee Member **Jon Arnot,** ARC Arnot Research & Consulting **Silvia Díaz-Cruz,** Institute of Environmental Assessment and Water Research **Michael Gonsior,** University of Maryland Center for Environmental Science **Bill Mitch,** Stanford University **Kurt Reynertson,** Johnson & Johnson Consumer Health
12:35–12:40	**Explanation of Breakout Session**
12:40–1:40	**Lunch Break (transition to breakout rooms)**
1:40–3:10	**Breakout Session on UV Filter Chemistry** *Format: Workshop participants will break into small groups to address the following questions:* 1. What are the main chemistry challenges encountered when working with (certain) UV filters? 2. Are challenges magnified when testing under certain conditions? 3. What progress is being made in addressing these challenges? 4. What standardizations, innovations, and/or other focused efforts are needed to move forward on addressing these challenges? 5. What are existing research programs, capabilities, and infrastructure that can contribute to addressing gaps in research on UV filter chemistry?
3:10–3:15	**Reconvene**
3:15–4:00	**Breakout Session Report Outs**

END OF DAY 1

TUESDAY, JANUARY 24, 2023

10:00–10:15	**Welcome and Review of Day 1** **Charles Menzie,** Exponent, Inc., Committee Chair

Session 2: Standardizing Approaches for Toxicity Testing

10:15–10:40	**The Importance of Standardized Toxicological Methods for Aquatic Organisms** **Sandy Raimondo**, Gulf Ecosystem Measurement and Modeling Division, U.S. Environmental Protection Agency
10:40–11:10	**Lightning talks: Methods for Coral Ecotoxicology** **Craig Downs,** Haereticus Environmental Laboratory **Sascha Pawlowski,** BASF **Abigail Renegar,** Nova Southeastern University
11:10–12:40	**Panel** *Format: Panelists will provide prepared remarks and then participate in discussion regarding progress, opportunities, and outstanding challenges relevant to aquatic toxicology of UV filters.* **Moderator: Carys Mitchelmore,** University of Maryland Center for Environmental Science, Committee Member **Mandy Annis,** U.S. Fish and Wildlife Service **Iain Davies,** Personal Care Products Council **Craig Downs,** Haereticus Environmental Laboratory **Marc Leonard,** L'Oreal **Sascha Pawlowski,** BASF **Sandy Raimondo,** U.S. Environmental Protection Agency **Abigail Renegar,** Nova Southeastern University **Jeffrey Steevens,** U.S. Geological Survey **Dan Villeneuve,** U.S. Environmental Protection Agency
12:40–12:45	**Explanation of Breakout Session**
12:45–1:45	**Lunch Break (transition to breakout rooms)**
1:45–3:15	**Breakout Session on Standardizing Approaches to UV Filter Toxicology** *Format: Participants will break into smaller groups to address the following questions:* 1. *What are the main challenges encountered when working with nonstandard organisms or endpoints?* 2. *Are challenges magnified when testing under certain conditions for both standard and nonstandard tests for these chemicals?* 3. *What progress is being made in addressing these challenges?* 4. *What standardizations, innovations, and/or other focused efforts are needed to move forward on addressing these challenges?* 5. *What are existing research programs, capabilities, and infrastructure that can contribute to addressing challenges in gaps in research on UV filter toxicity?*
3:15–3:20	**Reconvene**

3:20–4:05   **Breakout Session Report Outs**

Session 3: Closing

4:05–4:15   **Summary Remarks**
                     Charles Menzie, Exponent, Inc., Committee Chair

MEETING ADJOURNS

Appendix C
Biosketches for Workshop Planning Committee Members

Charles A. Menzie
Chair

Charles A. Menzie is principal and former practice director at Exponent, Inc. He was global executive director for the Society of Environmental Toxicology and Chemistry (SETAC) from 2014 to 2020. He specializes in the application of ecological and human health risk assessment and causal analysis methods for evaluating the potential for effects and for diagnosing the causes of environmental harms and damages. His technical expertise includes the evaluation of the environmental fate and effects of physical, biological, and chemical stressors on terrestrial and aquatic systems. He has applied his expertise to situations involving nutrient enrichment, chemical contamination, use of pesticides and other chemical products, oil and gas operations, fossil fuel and nuclear power plants, alternative energy projects, mining, invasive species, water management, and vulnerability assessments for climate change. As part of his risk assessment practice, he has developed exposure and food web models to evaluate how people and ecological receptors may be exposed to a variety of chemicals. These include several spatially explicit models used to refine exposure estimates. He previously served on the National Academies Committee on the Bioavailability of Contaminants in Soils and Sediments. Dr. Menzie has a B.S. in biology from Manhattan College and an M.A. and Ph.D. in biology from City University of New York.

Scott Belanger
Member

Scott Belanger retired as a research fellow from the Global Product Stewardship Global Capability Organization (Environmental Stewardship and Sustainability) of the Procter & Gamble Company, where he was in charge of aquatic toxicology research and led Procter & Gamble's overall environmental toxicology function. He was chair of the Corporate Function Safety Innovation Research Programs for both human and environmental safety. His research involves the response of aquatic organisms to single compounds and complex mixtures including classical ecotoxicological endpoints, bioaccumulation, and critical body burdens of test compounds. He is the founder and former chair of the HESI (Health and Environmental Sciences Institute) Project Committee on Animal Alternative Needs in Environmental Risk Assessment, a consortium of approximately 100 academic, industry, regulatory, and NGO scientists and presently serves as a member of the HESI Board of Trustees, a 501(c)(3) nonprofit. Dr. Belanger is a member of several ongoing Organisation for Economic Co-operation and Development (OECD) working groups, including the OECD Fish Framework, the OECD ad hoc Expert Group on the Fish Embryo Test (FET), and the Acute Fish Toxicity IATA working group. He was appointed SETAC (Society of Environmental Toxicology and Chemistry) Science Fellow in 2017. Dr. Belanger continues to work with regulatory agencies, academic units, and the private sector in environmental risk assessment matters including weight of evidence formulation, extrapolation of laboratory findings to the field, and use/interpretation of higher-tier forms of environmental scientific impact evidence. Dr. Belanger received a B.S. in zoology from the University of Wisconsin, a M.S. in biological sciences from Bowling Green State University, and a Ph.D. in zoology and aquatic ecotoxicology from Virginia Polytechnic Institute and State University.

Carys Mitchelmore
Member

Carys L. Mitchelmore is a professor at the University of Maryland Center for Environmental Science, Chesapeake Biological Laboratory in Solomons, Maryland. Her expertise is in environmental health and aquatic toxicology and her research emphasis is on understanding the exposure to, fate and effects of pollutants in resident organisms, particularly corals. Research is directed toward the detection of chemical contaminants in various environmental matrices and understanding their routes of exposure, uptake and bioaccumulation, metabolism, mechanisms of toxicity and implications to organism and ecosystem health. Applied research includes toxicity testing for application to risk assessment, regulation and management activities and providing solutions to applied environmental problems, such as, invasive species control. Recent investigations have focused on the chemical partitioning, fate and effects of organic UV filters, crude oils, oil spill dispersants and organic disinfection by-products in numerous invertebrate and vertebrate species, but especially sensitive and/or understudied species like corals and reptiles. Dr. Mitchelmore has received funding and travel support from the Personal Care Products Council for her research and testimony on UV filter effects on corals. She provided testimony to the Hawaii legislature regarding the ban of sale of certain sunscreen ingredients in Hawaii. Dr. Mitchelmore has served on two previous National Academies Studies: the Committee on the Effects of Diluted Bitumen on the Environment: A Comparative Study (2016) and the Committee on Understanding Oil Spill Dispersants: Efficacy and Effects (2005). She was also a review coordinator for the recent 2020 Committee on the Use of Dispersants in Marine Oil Spill Response. She currently serves on the National Academies Committee on Oil in the Sea IV: Input, Fates and Effects. Dr. Mitchelmore received her Ph.D. from the University of Birmingham (UK) in 1997 investigating the metabolism and effects of organic contaminants to aquatic organisms.

Robert Richmond
Member

Robert Richmond is the director and a research professor at the Kewalo Marine Laboratory at the University of Hawaii at Manoa. He received his doctorate in 1983 from the State University of New York at Stony Brook with a concentration in biological sciences. His research interests are focused on coral reef ecosystems, with studies including coral reproductive biology, ecotoxicology, coral reef ecology and the impacts of climate change. In 2006, he was awarded a Pew Fellowship in Marine Conservation during which he developed molecular biomarkers of stress in corals as a tool for coral reef conservation. In 2014, he received an award from the U.S. Coral Reef Task Force in recognition of advancing scientific research, mentoring and service. He was awarded grants from the Hawaii State Department of Health, NOAA and the National Fish and Wildlife Foundation to develop biomarkers of toxicant exposure in corals in Hawaii. Dr. Richmond has provided testimony to the Hawaii legislature regarding the ban of sale of certain sunscreen ingredients in Hawaii. Dr. Richmond is currently a member of the Palau International Coral Reef Center's board of directors and was a member of the Climate Change and Coral Reefs working group at the Center for Ocean Solutions. He is a past president of the International Coral Reef Society and served as the convener for the 13th International Coral Reef Symposium held in Hawaii in 2016. He previously served on the National Academies Committee on Interventions to Increase the Resilience of Coral Reefs.

Emma J. Rosi
Member

Emma J. Rosi is a senior scientist at the Cary Institute of Ecosystem Studies. Previously, Dr. Rosi was an assistant professor at Loyola University of Chicago. Dr. Rosi conducts research on factors that control and influence ecosystem function in aquatic ecosystems. Her research focuses on human modifications to freshwater ecosystems such as land use change and restoration, widespread agriculture, urbanization, the release of novel contaminants, and hydrologic modifications associated with large dams. Her research spans ecosystems from small streams to large rivers and has been conducted throughout the world, and includes biogeochemistry, secondary production, food webs, carbon cycling, and the effects of emerging contaminants on ecosystem processes. She is the director of the Baltimore Ecosystem Study Long-term Ecological Research Site. Currently she conducts research investigating how microplastics and pharmaceuticals affect stream ecology and food webs. She has served as an adviser to the EPA through service on committees of the Science Advisory Board, on the board of *Freshwater Biology*, as an associate editor for *Ecosystems*, as a reviewer for proposals submitted to NSF and USDA as well as for numerous scientific journals. She holds a Ph.D. and M.S. from the University of Georgia and a B.S. from the University of Michigan.

Cheryl Woodley
Member

Cheryl M. Woodley has served as a research scientist with the National Oceanic and Atmospheric Administration's National Ocean Service, National Centers for Coastal Ocean Science in Charleston, South Carolina for the last 30 years. With expertise in biochemistry, cellular biology, and pathobiology, she leads a multidisciplinary research team focused on understanding the effects of physical, chemical, and/or biological risk factors affecting conservation and management of vulnerable shallow-water coral species. Specific areas of interest include coral reproduction, disease pathologies and treatment, ecotoxicology, and diagnostic assay development. Dr. Woodley also holds an adjunct graduate faculty position at the College of Charleston and serves as the coordinator for the Coral Disease and Health Consortium, a working group of the U.S. Coral Reef Taskforce. She is coeditor of a comprehensive reference book (*Diseases of Coral*) and has coauthored more than 65 publications, two of those publications concern the toxicological effects of UV filters (benzophenone-2 and oxybenzone) on coral. Upon request and in her personal capacity, Dr. Woodley has provided expert opinions to several city, state, and federal legislators on the impacts of ultraviolet filters on coral health. Dr. Woodley completed her doctorate in the molecular, cellular, and pathobiology program at the Medical University of South Carolina, studying serine proteases in the kallikrein–kinin system and received specialized training in virology at Baylor College of Medicine, Houston.